Modeling and Optimization of Solar Thermal Systems:

Emerging Research and Opportunities

Jagadish
National Institute of Technology, Raipur, India

Agnimitra Biswas
National Institute of Technology, Silchar, India

A volume in the Advances in Mechatronics and Mechanical Engineering (AMME) Book Series

IGI Global
PUBLISHER of TIMELY KNOWLEDGE

Published in the United States of America by
IGI Global
Engineering Science Reference (an imprint of IGI Global)
701 E. Chocolate Avenue
Hershey PA, USA 17033
Tel: 717-533-8845
Fax: 717-533-8661
E-mail: cust@igi-global.com
Web site: http://www.igi-global.com

Library of Congress Cataloging-in-Publication Data

Names: Jagadish, 1986- author. | Biswas, Agnimitra, 1980- author.
Title: Modeling and optimization of solar thermal systems : emerging research and opportunities / by Jagadish and Agnimitra Biswas.
Description: Hershey, PA : Engineering Science Reference, an imprint of IGI Global, [2020] | Includes bibliographical references and index. | Summary: "This book investigates solar thermal systems using multiple-criteria decision analysis. It also explores recent advancements in efficiency, size miniaturization, cost effectiveness of solar technologies, and performance optimization using multiple-criteria decision analysis"-- Provided by publisher.
Identifiers: LCCN 2019054347 (print) | LCCN 2019054348 (ebook) | ISBN 9781799835233 (hardcover) | ISBN 9781799835240 (paperback) | ISBN 9781799835257 (ebook)
Subjects: LCSH: Solar thermal energy--Mathematical models. | Solar collectors--Design and construction.
Classification: LCC TJ810 .B488 2020 (print) | LCC TJ810 (ebook) | DDC 621.47--dc23
LC record available at https://lccn.loc.gov/2019054347
LC ebook record available at https://lccn.loc.gov/2019054348

This book is published in the IGI Global book series Advances in Mechatronics and Mechanical Engineering (AMME) (ISSN: 2328-8205; eISSN: 2328-823X)

British Cataloguing in Publication Data
A Cataloguing in Publication record for this book is available from the British Library.

Advances in Mechatronics and Mechanical Engineering (AMME) Book Series

ISSN:2328-8205
EISSN:2328-823X

Editor-in-Chief: J. Paulo Davim, University of Aveiro, Portugal

MISSION

With its aid in the creation of smartphones, cars, medical imaging devices, and manufacturing tools, the mechatronics engineering field is in high demand. Mechatronics aims to combine the principles of mechanical, computer, and electrical engineering together to bridge the gap of communication between the different disciplines.

The **Advances in Mechatronics and Mechanical Engineering (AMME) Book Series** provides innovative research and practical developments in the field of mechatronics and mechanical engineering. This series covers a wide variety of application areas in electrical engineering, mechanical engineering, computer and software engineering; essential for academics, practitioners, researchers, and industry leaders.

COVERAGE

- Sustainable and green manufacturing
- Design and Manufacture
- Computational Mechanics
- Intelligent Navigation
- Mechanisms and machines
- Tribology and surface engineering
- Biologically Inspired Robotics
- Bioengineering Materials
- Nanomaterials and nanomanufacturing
- Medical robotics

IGI Global is currently accepting manuscripts for publication within this series. To submit a proposal for a volume in this series, please contact our Acquisition Editors at Acquisitions@igi-global.com or visit: http://www.igi-global.com/publish/.

The Advances in Mechatronics and Mechanical Engineering (AMME) Book Series (ISSN 2328-8205) is published by IGI Global, 701 E. Chocolate Avenue, Hershey, PA 17033-1240, USA, www.igi-global.com. This series is composed of titles available for purchase individually; each title is edited to be contextually exclusive from any other title within the series. For pricing and ordering information please visit http://www.igi-global.com/book-series/advances-mechatronics-mechanical-engineering/73808. Postmaster: Send all address changes to above address.

Titles in this Series

For a list of additional titles in this series, please visit:
http://www.igi-global.com/book-series/advances-mechatronics-mechanical-engineering/73808

Evaluation and Applications for Ultrasonic Methods of Non-Destructive Testing
Ashish Khaira (Samrat Ashok Technological Institute, India)
Engineering Science Reference • © 2021 • 215pp • H/C (ISBN: 9781799845645) • US $195.00

Handbook of Research on Advancements in Manufacturing, Materials, and Mechanical Egineering
Leonid Burstein (Independent Researcher, Israel)
Engineering Science Reference • © 2021 • 462pp • H/C (ISBN: 9781799849391) • US $295.00

Genetic Algorithms and Remote Sensing Technology for Tracking Flight Debris
Maged Marghany (Syiah Kuala University, Indonesia)
Engineering Science Reference • © 2020 • 334pp • H/C (ISBN: 9781799819202) • US $195.00

Humanizing Cities Through Car-Free City Development and Transformation
Rahma M. Doheim (Assiut University, Egypt & University of Business and Technology, Saudi Arabia) Alshimaa Aboelmakarem Farag (Zagazig University, Egypt) and Ehab Kamel (University of Central Lancashire, UK)
Engineering Science Reference • © 2020 • 305pp • H/C (ISBN: 9781799835073) • US $215.00

Airline Green Operations Strategies Emerging Research and Opportunities
Yazan Khalid Abed-Allah Migdadi (Qatar University, Qatar)
Engineering Science Reference • © 2020 • 231pp • H/C (ISBN: 9781799842552) • US $185.00

For an entire list of titles in this series, please visit:
http://www.igi-global.com/book-series/advances-mechatronics-mechanical-engineering/73808

IGI Global
PUBLISHER of TIMELY KNOWLEDGE

701 East Chocolate Avenue, Hershey, PA 17033, USA
Tel: 717-533-8845 x100 • Fax: 717-533-8661
E-Mail: cust@igi-global.com • www.igi-global.com

Table of Contents

Preface

Solar energy research is not a new thing. It is very much pervasive over decades from the last century. Solar energy is the most prolific form of renewable energy on earth. In the present era when there is a dearth of conventional energy as well as the danger of burning fossil fuels being so much of a concern, harvesting renewable energy like solar energy is the way forward for fulfilling the energy requirements of human and industry. Solar energy is harvested by using solar energy systems and devices. Solar energy systems are broadly classified into two types: Solar Thermal Systems and Solar Photovoltaic Systems. Further, there is hybrid solar energy system as well. While the first type is dedicated for producing solar thermal energy, the second type is dedicated for producing solar power. The hybrid system can generate energy in different forms. For effective utilization of solar energy, researchers and scientists are in continuous quest for improving the design and configuration of solar energy harvesting systems. Whereas, there is always an ardent need for modeling and optimization of solar energy systems for improving the degree of accomplishment of the objective, i.e. effective harvesting of solar energy.

THE CHALLENGE

As mentioned above, the challenge in solar energy harvesting is to improve the effectiveness of solar energy systems. The effectiveness depends on various design and operating aspects. These do not only depend on sophisticated designs of solar energy systems but also depend on how best the operating conditions can be optimized for the proper functioning of the system. Although various scientists have attempted to improve the performances of solar energy systems through design innovation and optimization, there still remains a major challenge as to how to select the best combination of operating conditions that guide the designers of such systems in ensuring effective performance of the solar energy systems.

SEARCHING FOR SOLUTION

When faced with the above challenge to ensure effective performance of the solar energy system, there is a scope for research opportunity in improving the same. There is a consortium of research methodologies that have been adopted to address this issue. However, not all methodologies are able to solve some specific problems associated with the solar energy systems. For example, intermittency of ambient conditions like solar radiation, ambient temperature, wind speed, etc. modifies performance of the system. The other operating conditions, for example mass flow rate, solar collector tilt angle, fluid inlet temperature can only be adapted to the changing ambient conditions. The earlier research works have used the ambient conditions under controlled solar environment (by using sun simulator) thereby facing less challenge of intermittency in improving the performance of the solar energy systems. However as the operating conditions are dynamic in nature, conventional approach of performance optimisation is not an easy solution of the problem. For example, conventional methodology like parametric optimization with having a control on the variables cannot result in an effective system design as there is variability in the ambient conditions and there are conflicting requirements of parameter variations as well. Hence, an emerging technique like multi criteria decision making (MCDM) approach can be tuned in the modeling and optimization of the various solar energy systems under dynamic conditions of all operating parameters. MCDM is a famous tool to address all sorts of uncertainties in the variations of parameters having conflicting requirements and even so when there are dynamic parameters like ambient parameters. The present book addresses different tools of MCDM technique for the modeling and optimization of both the categories of solar energy systems, i.e. solar thermal energy systems and solar PV systems. The book also addresses the problems associated in the modeling of hybrid solar energy systems under varying and conflicting conditions of performance and cost.

The present book contains six chapters starting with an introduction to solar energy systems followed by its classifications and modeling and optimization issues of various thermal systems using MCDM method with detailed summary are discussed in chapter 1. Chapter 2 presents the modeling and optimization of SFPC system parameters for ideal parameters such as outlet temperature, efficiency, useful heat gain, exergetic efficiency using modified-FST-MOOSRA method. Also, for understanding the effectiveness of the model results with optimized parametric conditions, analysis of variance and also a confirmatory test are conducted, and finally the summary is drawn at the end. In chapter 3 presented the application of DEAR method in modeling and optimization of CST system to improve the optimal response like temperature, height of the receiver followed by ANOVA and parametric analysis and confirmatory tests to verify the results further. Chapter 4 reports on implementation of Modified SAW method for modeling and optimization of ETSTC system followed by some statistical analysis (parametric analysis, ANOVA, and empherical models) to verify the results further. Implementation of integrated method i.e. Entropy with GTMA for modelling and optimization of PTC parameters to improve higher outlet temperature (T_o), higher heat gain and higher thermal efficiency is discussed in chapter 5. In last chapter 6 discussed the on modeling and optimization of hybrid system based solar PV System with Pumped Hydro Energy Storage system followed by application of integrated method i.e. AHP-MOORA to progress in performance of PV-PHS system along with ANOVA and confirmatory tests .

Authors hope that the work reported in this book would help and inspire the engineers, researchers, and experts working in similar field.

We really acknowledge IGI global for this probability and their expert support.

Jagadish
National Institute of Technology, Raipur, India

Agnimitra Biswas
National Institute of Technology, Silchar, India

Chapter 1
Introduction

ABSTRACT

This chapter introduces to the solar energy in general and in perspective as far as the applications and utility of solar energy systems are concerned. It gives the brief overviews of all types of solar energy systems starting with their classifications followed by their modeling and optimization issues. Work sheds light on both solar thermal systems and solar photovoltaic (PV) systems. The performance measurement of these solar energy systems strongly depends on various system and operating parameters along with their ambient or climatic parameters, and it is a multi-criteria decision-making (MCDM) optimization problem. Based on the previous studies, the statistics as regards to the modeling and optimization of different solar energy systems and the basics of MCDM techniques with their classifications are discussed. The chapter ends with detailed summary with possible scope of future work on modelling and optimization of solar energy systems using various individuals plus hybrid MCDM techniques.

INTRODUCTION TO SOLAR ENERGY AND ITS SYSTEMS

Solar energy is the most prolific form of renewable energy resource that is being harnessed by the mankind since many years. Solar energy is received at the earth surface due to the result of millions of fusion reactions that take place at the core of the Sun. The Sun is a giant sphere with a diameter 1.39 x 10^9m, which generates intensely hot gaseous matter through fusion reactions in which hydrogen is turned into helium (Duffie and Beckman, 1980). The

DOI: 10.4018/978-1-7998-3523-3.ch001

sun subtends an angle of 0.53^0 on the earth surface and the solar energy strikes our planet in 8 min and 20 s after leaving the Sun's surface, which is 1.495×10^{11} m away from the earth (Duffie and Beckman, 1980). The solar radiation is created from the outer convective zone of the Sun, which has an effective blackbody temperature of 5762K. However, the temperature in the central region of the Sun is much higher, which is estimated as 8×10^6 to 40×10^6 K. The Sun's total energy output is 3.8×10^{20} MW, which is equal to 63 MW/m^2 of the Sun's surface which radiates outwards in all directions. However, only a tiny fraction, 1.7×10^{14} kW of the total radiation emitted, is intercepted by the earth. It is estimated that 30 minutes of this small fraction of solar radiation falling on the earth is equal to the world energy demand for one year. The world is bestowed with good and usable solar insolation level that can be harnessed effectively for the benefit of the society (Sukhatme and Nayak, 2017; Kreith et al. 1978).

In the scarcity of conventional fuel and the increased price level along with the danger of environmental degradation, renewable energy has tremendous implications in mitigating the crisis and satisfying the energy requirements of the people. The handbook of research on renewable energy and electric resources for sustainable rural development (Kharchenko & Vasant, 2018) explores the effective use of renewable energies with a positive impact on sustainable development. The primary topic includes renewable energy systems application practices, cogeneration thermal modules, photovoltaic solar, electric pulse cultivation, and thermal treatment. Generation and delivery of renewable energy in rural area is a difficult task. This requires special technical supports. The handbook, renewable energy and power supply challenges for rural regions (Kharchenko & Vasant, 2019) provide innovative perspectives into the growth, use and delivery of energy in rural areas and also provide the analysis of alternative and renewable energy sources. The handbook explores such topics as renewable energy, electrical networks, thermal energy storage, electric transport, solar modules, etc. The handbook, research on smart computing for renewable energy and agro-engineering (Kharchenko & Vasant, 2020) explores the advantages and challenges of incorporating agricultural and renewable energy production models, and how these models can produce cost-effective and sustainable solutions. The handbook of research on energy-saving technologies for environment friendly agricultural development (Kharchenko & Vasant, 2020) offers insights into emerging developments that build productive and environmental-friendly energy capacity needed to mobilize rural and industrial economy. Handbook stresses on topics including economic development, renewable energy, climate change, harvesting machinery, power supply system etc. Babaev et al. (2019) performed a detail analysis on types of accumulating energy and the classification of the heat accumulation parameters in energy system depending on the material class, how the heat is stored and how the work is cyclical.

From the perspective of energy and sustainable economic development, renewable energy like solar energy has evolved as one of the primary alternatives to which further efforts are being directed to satisfy current energy needs without compromising future generations. Solar energy has diversified application areas ranging from solar photovoltaic power, solar thermal power, hybrid solar energy, solar absorption refrigeration, solar passive heating and cooling, polygeneration, and many more. In this book, various solar thermal systems as well as solar PV systems have been discussed. Solar PV systems have the unique capability to produce electric power as well as thermal energy since the heat absorbed by the PV panel is a source of heat, which can be taken up by a coolant in a secondary circuit for generating thermal energy that can be used in various solar thermal applications. Solar thermal systems can generate heat energy and can also produce solar thermal power. That is to say a solar thermal system transforms solar energy into heat energy in the form of hot fluid, which can be used in its direct form or can be further harnessed to produce solar thermal power. Groenendaal (2002) discussed the status of various technologies for solar thermal power (STP). Sychov et al. (2018) used different computer tools for optimizing a residential water-to-air heating and air conditioning system with low-grade heat. The heart of a solar thermal system is called solar collector. Different solar collectors have been developed by the scientists and researchers over many years. Solar collectors are normally of two types- flat plate collector (mainly for heating) and concentrated (line and point focussing) collector (for solar thermal power generation). Solar thermal system is capable of producing solar thermal power in the scale of megawatts. Solar thermal systems for power generation are of two types- solar aided system and solar alone system. In solar aided system, solar energy is used along with the primary fuel for heating in order to produce steam or hot gaseous mixture to generate electric power, for example solar assisted Rankine cycle power plant, solar assisted Brayton cycle power plant, solar assisted gas fired combined cycle power plant, etc. However, in case of solar alone system, a concentrated solar collector replaces the boiler or the combustion chamber to generate heat or gases to operate on the turbines, for example solar Rankine cycle power plant, solar Organic Rankine cycle (ORC) power plant, central receiver system solar thermal plant, Paraboloid dish concentrating solar thermal power plant (Stirling engine), etc. In solar thermal area, solar energy is being used in variety of day-to-day domestic and industrial uses, for e.g. steam cooking, water heating, space heating, drying, desalination, distillation, absorption refrigeration, solar cooling, solar air conditioning, polygeneration, concentrated solar thermal power generation, etc. (Kalogirou, 2004).

A solar PV system transforms solar radiant flux into usable form of solar energy i.e. electric power. Solar PV system is capable of generating electric power starting from household requirement to mega-scale electric power requirement. The photovoltaic effect was first observed by French physicist Alexandre-Edmond Becquerel in 1839. The heart of the solar PV system is its PV panel, which is made from semi-conductor materials called solar cells. The most common material used in solar cells is silicon. Solar PV panel is made from modules that consist of silicon cells. As solar radiation falls on the solar cells, it energies the electrons that migrate from the valence band to the conduction band thereby creating an electromotive force (emf), which is the result of electric current. Each solar cell is capable to produce 0.5V and by connecting these solar cells in series and parallel combination, the necessary DC voltage is obtained from the PV panel. There are different types of PV cells- monocrystalline, polycrystalline, and thin film. Out of these thin film solar cell has the highest cell efficiency (around 21% in laboratory setting). A complete solar PV system is created when the PV panel is integrated with an inverter and an electric energy storage system, and other balance of systems. The handbook of research on solar energy systems and technologies (Anwar et al. 2013) report the use of nanotechnology in solar power research including the current and future advances of solar cells of third generation.

Applications of Solar Thermal Systems and Solar PV Systems

Solar thermal systems find wide applications in households, and also in industries. The description of various solar collectors with indicative temperature range and application is given in Table 1.

Solar PV systems are used in varieties of applications and these are of various forms. The major use of a solar PV system is solar power generation (both DC and AC) by using solar PV plant, just like thermal power plants. In addition, with the help of PV panel, solar energy is also used in solar street lights, solar DC pumping unit for irrigation, solar agro sprayer, solar lanterns, solar home lighting system; similarly in other different solar PV gadgets like

Table 1. Descriptions of variety of solar collectors

Solar Collector	Types	Absorber type	Indicative temperature range	Application
Non-Concentrating collector	Flat Plate Collector (FPC)	Flat	30°-80°C	Water heating, drying, space heating, distillation
	Evacuated Tube Collector (ETC)	Flat	50°-200°C	Water heating, Process heating, pasteurizing
	Compound Parabolic Collector (CPC)	Tubular	60°-240°C	Water heating, Process heating, air-conditioning
Concentrating collector				
	Linear Fresnel Reflector (LFR)	Tubular	60°-300°C	Heating, steam power generation
	Parabolic Trough Collector (PTC)	Tubular	60°-400°C	Heating, steam power generation
	Parabolic Dish Reflector (PDR)	Point	600°-900°C	Community Cooking, steam power generation
	Heliostat Field Collector (HFC)	Point	600°-2000°C	-do-

solar calculator, portable solar charger, solar projector, solar backpacks, solar torch, solar watches, solar headset, solar cameras, etc. Apart from that solar PV panels are also being used for running automobiles by charging batteries externally, solar racing cars, solar rickshaws, flying aeroplanes and drones, powering satellite by mounting solar panels on its body, etc., although most of these are still in research stages.

Advantages and Limitations of Solar Energy Systems

There are more advantages than disadvantages of renewable energy systems for those matter solar thermal systems. Solar thermal systems have so many advantages, for e.g. can generate clean form of energy, i.e. no greenhouse gas emissions, inexhaustible form of energy source from Sun's radiant flux, capability to replace fossil fuel based energy systems, having large number of domestic and industrial applications, possibility of energy conversion into the form of solar thermal power, noise-free, have ecstatic and customized designs, etc. There are a few disadvantages of solar thermal systems, for e.g. high initial investment cost, less efficient compared to fossil fuel based energy systems, require energy storage unit for storing thermal energy during Sun-less periods, possibility of greenhouse gas emissions during the manufacture of the components, etc.

The solar PV systems have gained more popularity than the solar thermal systems due to the more flexibility of storing energy in electrical energy storage systems like battery bank, fuel cell, ultra capacitor, etc. However, storage of solar thermal energy is a big challenge as the heat is a boundary phenomenon as well as a transient phenomenon, and for which it is difficult to store. Solar PV system is a mature technology and has become commercially more successful then the solar thermal system. Solar PV system reduces transmission and distribution losses since power is consumed at the point of generation. Both DC and AC power can be generated by such system. The major issue facing the solar PV systems is the maintenance of the solar PV plant, short battery life, and the initial investment cost. In addition, the cell efficiency reduces as they get heated up due to the incident solar radiation. Nevertheless both solar thermal and solar PV systems have their importance in harnessing solar energy for mankind in varieties of applications. Now-a-days more advancement has been made in the storage of solar thermal energy in the form of bulk storing by phase-changing with the help of Phase Change materials (PCM), storing by the using the combination of nanoparticles and PCM.

Classification of Solar Energy Systems

A broad classification of solar thermal systems is represented in Figure.1. The heart of the solar thermal system is its thermal collector that captures the solar radiation for further processing. The thermal collectors are of two types- concentrating (non-stationary) and non-concentrating (stationary), which have further sub-classifications as shown in Figure.1.

Solar thermal technologies can also be classified on the basis of output temperature of the working fluid. It would be called low temperature technology when output temperature is of the order of 100°C; called medium temperature technology when output temperature is of the order of 100 °C-300°C, and high temperature technology for output temperature above 300°C. For these temperature ranges, solar thermal technologies are used for a host of household and industrial applications. Avezov et al. (2018) studied the flat plate collectors and obtained the dependence of flat-plate solar water heating technologies and economic indicators on the water heating system temperature. Klychev et al. (2018) showed how a thermal modal can be used to determine the temperature of solar flat plate water collector.

Figure 1. Classification of solar thermal systems

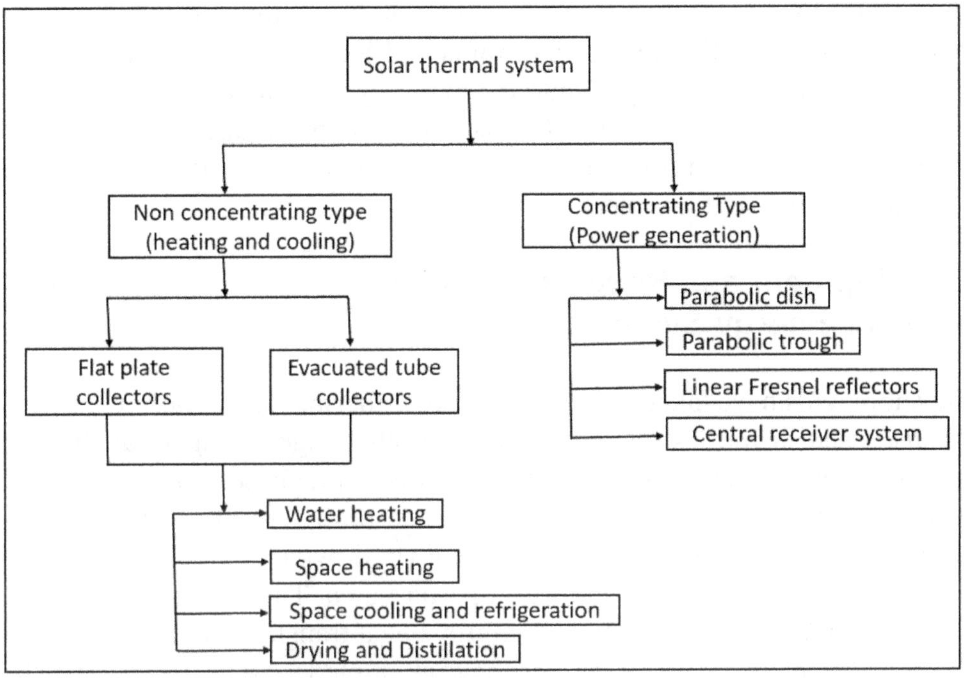

Solar PV systems are of different designs, which are classified according to the following categories- off-grid solar PV system with battery bank, grid-connected or grid-tied solar PV system without battery bank, and hybrid solar PV system i.e. combination of grid-tied and off-grid solar PV system. The off-grid solar PV systems, called standalone solar PV system, are classified into two types- DC off-grid system and AC off-grid system. An off-grid solar PV system normally includes PV panels, charge controller, battery bank, and electric load for DC power. If AC power is required then a DC to AC converter, i.e. an inverter is used. In the off-grid systems, the solar power is first stored in the battery bank after drawing current through the charge controller in the charging mode. The output of the PV panel is supplied to the charge controller. Then in the discharging mode, the current is drawn from the battery into the controller and then the electric power is transferred from the controller to the DC load or to the AC load after DC to AC conversion by using an inverter. The grid-tied PV system includes the solar PV plant and the utility grid that is interconnected by using a bidirectional interface called a net metering unit. The consumers pay for the consumption recorded in the net metering unit if electricity is drawn from the grid. The solar energy generated by the PV panels is used for internal use of the households, and the electricity (if produced) is transferred to the utility grid through the net metering unit. Similarly during rains and Sun-less periods, the electric current is drawn from the grid into the metering unit and then transferred to the households for internal use. In this book, a configuration of an off-grid solar PV system with a unique storage system is presented and discussed (included in the sixth chapter).

Development and Research Issues in Solar Energy Systems

Developments have taken place in both solar thermal and solar PV systems as discussed in the last section. The main challenge facing solar thermal systems is its low efficiency compared to fossil-fuel based energy systems. The contributing factor to low efficiency is the result of inefficient heat transfer from the collector to the working fluid, which also involves lot of thermal losses due to large collector aperture area. Thus, the research issue is how to minimize the losses and improve heat transfer rate to enhance the system efficiency. Many researchers have attempted to solve this problem by variety of ways, for e.g. using better heat transfer fluids like nanofluids, using high heat conducting absorber surface like absorber made of Cu, improved insulations to lower heat losses, augmentation of solar insolation falling on the top of collector by concentration, etc. (Poulek, 2010). Another way to control losses is by restricting the temperature of the collector working fluid upto a certain level for which losses due to finite temperature difference can be lowered. In addition to that the thermal efficiency of the collector also depends on the temperature difference between outlet and inlet of the collector; two much temperature difference leads to lowering of efficiency due to increased losses from the collector. Another issue is the variable climatic conditions that impact the performance of the collector in addition to the impacts of collector design and various operating and ambient conditions, for example mass flow rate of working fluid, ambient temperature, and intensity of solar radiation. Therefore, although the solar thermal collectors have been developed since their inventions that happened in the mid-19th century, still their designs have variable performances and the same is location specific as well. The reason for this is the lack of optimization of its design and operating parameters under variable weather conditions (i.e. variable solar radiation, ambient temperature, wind speed, humidity condition). Further, the design and operating parameters together have conflicting influences on the performance of the systems, which require an MCDM modeling of the thermal systems, which is still lacking in the available literature.

The major research issue in solar PV system is the technical and economic design of the system considering the application area. Solar PV systems have various losses for not using the system optimally, for example deficit of energy (i.e. lack of reliability) or excess energy generated by the system, which largely affect the cost of the system. The cost of energy is one such parameter, which depends on the size and cost of the system components in addition to the operating conditions like solar radiation and ambient temperature. High ambient temperature increases the cell temperature thereby dropping the PV cell efficiency. Hence the PV capacity design with optimum operating conditions is required to make a successful solar PV system. Thus, there can be multiple output and input parameters to be considered in the performance optimization of solar thermal and solar PV systems, which have conflicting requirements. Further, variable operating conditions like excess energy, deficit of energy etc. directly influences the cost of the system and performance of the solar PV system, i.e. the cost of the energy. The cost of energy is also influenced by the design parameters, i.e. size of the PV system and the size of the storage units, which are also variable depending upon the capacity required. All these variable input and output parameters can be entailed and the performance of the system can be optimized by adopting the multi-dimensional framework of MCDM techniques.

Modeling and Optimization of Solar Thermal Systems

The thermal performance of solar thermal systems is dependent on various design and operating parameters including ambient or climatic parameters. The design parameters include collector size, collector tilt angle, altitude angle, incidence angle, etc. The major operating parameters are mass flow rate through the collector and also fluid inlet temperature to the collector. The ambient parameters are solar radiation, ambient temperature, humidity, etc. The solar collector is the heart of the solar thermal system. The performance of the collector is largely influenced by the composite climatic conditions of the place where the solar thermal system is installed. As the climatic conditions are different in different parts of the world, performance of the same solar thermal collector is different in different places. In addition performance of solar thermal collector is largely dependent on its design and operating parameters. There are desirable ranges of operating parameters in which performance of the collector will be higher. However, the design conditions should also be effectively chosen to obtain the desirable efficiency of the collector. Thus, there must be tailor-made designs of solar thermal collectors based on region-specific applications so that the performance of the collector can be optimized.

Given the multitudes of conditions (design, operating, and climatic) to be fulfilled, a detailed parametric analysis of the collector performance can be undertaken to evaluate the influence of the above conditions on its performance. However, as all these conditions have conflicting requirements, hence the decision-making about the selection of suitable design and operating parameters for improved performance of the solar thermal collector is also a complex task. The issues related to conflicting requirements for solar thermal systems for e.g. various design and operating parameters have conflicting importance in the performance improvement of the solar thermal systems. As an example, in order to harness more and more solar energy using flat plate collector, its surface area can be increased but the same results in increased thermal losses from its top glazing. The losses in turn are strongly influenced by the intensity of solar radiation and the ambient temperature, which themselves are variable. Thus performance of thermal collectors is largely affected by variable operating conditions and the design choice. To handle such conflicts, the present book presents to MCDM approaches to model the performance of solar thermal collectors. One of the major advantages of MCDM is its ability to quantify the relative importance of the variable operating and design parameters in terms of their influences on the performance optimization of the systems, which conventional techniques cannot do so. Moreover, statistical significance of all such parameters for sustainable use of the solar thermal collectors is also evaluated for each design choice and the relevant chapters have also identified the beneficial (significant) and non-beneficial (insignificant) parameter choices for effective use of the solar thermal collectors.

Although the solar thermal collectors have been developed since their inventions that happened in the mid-19th century, still their designs have variable performances and the same is location specific as well as also discussed above. The reason for this is the lack of optimization of its design and operating parameters under variable weather conditions (i.e. solar radiation, ambient temperature, humidity condition). As the design and operating parameters have conflicting influences on the performance of the systems, it requires an MCDM modeling of the thermal systems, which is still lacking in the available literature. A simple parametric analysis may not always find optimum conditions for maximum performance of the collector whenever there is more than one output parameter. Moreover, the selected design and operating parameters should also comply with the constraints of the climatic conditions of the place where the collector has to function. All these situations entail the application of MCDM in the modeling and performance optimization of the solar thermal systems. Further, different MCDM techniques may be applied for modeling and performance optimization of the solar thermal systems to find an effective MCDM technique for a particular system.

Introduction to MCDM Techniques

The use of multi criteria analysis for providing solutions to complex energy management problems has attracted the attention of decision makers for quite some time now. These

methods provide better understanding of the intrinsic features of a decision problem by making them more explicit, efficient and rational as opposed to traditional single criteria decision making, which are normally aimed at maximization of performance benefits or minimization of costs. MCDM is a sub-discipline of operations research that is widely used in many fields such as energy planning, construction, environmental projects, software engineering, military applications, food safety, solid waste management, agricultural production, risk management, manufacturing systems (Mardani et al. 2015). Multi criteria analysis promote the role of participants, facilitate compromise and collective decisions thus providing a good platform in understanding the perception of models and analysts in a realistic scenario. MCDM methods can be divided into Multi-Objective Decision Making (MODM) and Multi-Attribute Decision Making (MADM) methods (Hwang and Yoon, 2012). They are further classified based on priority, outranking, distance, etc. Each method has its own characteristics and the methods can also be classified as deterministic, stochastic and fuzzy methods. There may also be combinations of the above methods. Depending upon the number of decision makers, the methods can be classified as single or group decision making methods. Decision making under uncertainty and decision support systems are also prominent decision-making techniques (Gal et al., 2013).

Decision-making problems consist of multiple numbers of criteria and these criteria are generally conflicting in nature. In order to make an effective decision, all the relevant criteria must be considered which has given various

MCDM techniques. These MCDM techniques usually consist of three basic stages viz. identification of the alternatives and criterions, criterion weight calculation and selection of the best alternative. In these techniques, a structured methodology is utilized to make the best decision out of all the available alternatives. The decision making process consists of mainly 6 steps, which are shown in Figure 2 (Jackson and Pascual, 2008). The criteria selected must be unique, must have the ability to be represented on a same scale and must clearly reflect the characteristics of all the alternatives. The selected alternatives must be comparable, available, real and not ideal, and must be practical. The choice of decision making tool depends on the type of problem (Sabaei et al. 2015). Finally, the results are aggregated using product, average or a function. These techniques find applications in various sectors such as maintenance management (Sabaei et al. 2015), life cycle analysis (Campos et al. 2019), automotive industries (Castro and Parreiras,2018), location selection for hybrid renewable energy systems (Diemuodeke, 2018), marine engineering sector (Jeong et al. 2014) and many more. There are different types of MCDM techniques, resulted from the applied researches of scientists over decades. Some of the popular MCDM techniques, which are dealt with in the present work are Multi-Objective Optimization on the basis of Simple Ratio Analysis (MOOSRA), Data Envelopment Analysis based Ranking (DEAR), Simple Additive Weighting (SAW), Fuzzy Set theory (FST), Multi-Objective Optimization on the basis of Ratio Analysis (MOORA) and Graph Theory and Matrix Approach (GTMA).

Figure 2. Steps of decision making

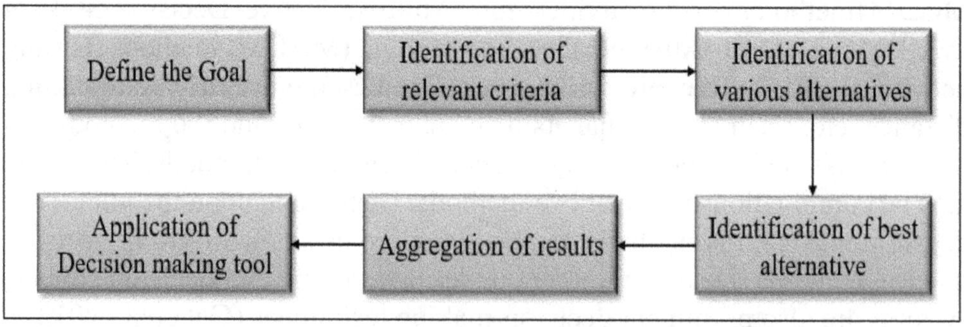

MCDM Techniques for Flat Plate Solar Thermal Collector

Real world problems constitute more than one criterion for which the multi criteria decision making approaches have been developed to solve such complex ordeals. The MCDM evaluates the overall preference among the existing alternative sets or options which may be considered closest to the ideal solution and furthest to the worst. Below is some of the literature on MCDM techniques applied for heat transfer augmentation in flat plate or rectangular channel based solar thermal collectors. Chauhan et al. (2016) used preference selection index (PSI) method for evaluating best alternative set for parametric selection in solar thermal collectors provided with impinging air jets. They evaluated total 18 alternative sets and optimized the best alternative combination using PSI method. Chamoli [26] used the same methodology for optimizing parameters in flat plate solar collectors with V down baffles as turbulence promoters. Chamoli, (2015) also studied the hybrid fuzzy Analytic Hierarchy Process (AHP) and Technique for Order of Preference by Similarity to Ideal Solution (TOPSIS) methodology for optimizing V down baffle parameters in a rectangular solar thermal collector having fluid flow passage. Sharma et al. (2017) optimized the baffle parameters of a solar flat plate air collector using multi criteria decision making in which weight determination was carried out using entropy method. An energy and exergy analysis of a novel solar air heating collector design was performed by optimizing the obstacle distances, obstacle relief angles, and Reynolds number using Grey Relational Analysis (GRA) method (Acır et al. 2017). GRA was integrated with Taguchi technique to optimize the processing parameters of a flat-plate thermal collector, which were collector tube material, endothermic plate material, number of collector tubes, collector tube diameter, absorption film type, and thickness of the bottom heat insulating material (Kuo et al. 2011).

Bouy et al. (2015) applied Preference Ranking Organization Method for Enrichment of Evaluations (PROMETHEE) for comparing two solar water heating technologies based on FPC and ETC in order to link with the climatic conditions of Morocco. Various issues were considered when comparing the technologies of collectors (FPC and ETC), in particular, economic efficiency, climatic impacts, etc. The potential of each technology to achieve these goals could be evaluated using different criteria. So, the MCDM methods seemed to be an appropriate choice for comparing FPC and ETC technologies. These methods could handle both quantitative as well as qualitative criteria and analyse conflict in criteria and decision makers. In particular, the most popular class of MADM was used to solve the problem. This class contains several methods: AHP, PROMETHEE, Elimination and Choice Expressing Reality (ELECTRE), Multiple Attribute Utility Theory (MAUT), etc. Following real life observations, the decision makers are often not fully aware of their preferences, or that they are not able to express these in unambiguous way. Thus, the PROMETHEE method was selected here as a methodological basis which provided an appropriate support to the decision makers. Sharma et al. (2017) to investigate the performance of discrete V obstacles in a solar air channel and the results were presented in terms of Nusselt number, friction factor and effective efficiency for thermal hydraulic and collector efficiency respectively. Since, the overall performance of solar air channel is governed by simultaneous consideration of all the three criteria, an optimization based upon Entropy-VIKOR (VIseKriterijumska Optimizacija I Kompromisno Resenje) approach has been reported which evaluates simultaneously these performance criteria. Where entropy measures the weights of each of the performance evaluation criterion towards overall best alternative set, the VIKOR method provides ranking of the alternatives by providing the best compromise solution which is nearest to the real and farthest from the worst.

MCDM Techniques for Concentrated Solar Thermal System

There are now several Concentrated Solar Thermal (CST) technologies, such as parabolic troughs and central receivers using various heat transfer media that have reached the phase of initial commercial deployment for large-scale power production. Information about cost, performance and impact on the environment of the systems being analysed is currently affected by uncertainty. In many cases, traditional evaluation methods such as cost-benefit analysis and the main economic and financial indicators are unable to deal with all the components involved in a valid energy project. Multi-criteria methods provide a flexible tool that is able to handle and bring together a wide range of variables appraised in different ways and thus offer useful assistance to the decision maker in mapping out the problem. Multi-criteria analysis can provide a technical-scientific decision making support tool that is able to justify its choices clearly and consistently, especially in the renewable energy sector.

Cavallaro (2010) in his study proposed and validated the effectiveness of TOPSIS fuzzy method by comparing different heat transfer fluids in a solar thermal energy storage system for a CST power system. Cavallaro (2009) again computed and displayed two distinct rankings of alternatives to assess the performance of concentrated solar thermal technologies. The first is PROMETHEE I, which gave a partial ranking. It was based on strongly established preferences so some actions remained incomparable under this method. This was followed by PROMETHEE II which gave a complete ranking: all actions were ranked from the best to worst leaving no incomparable pair of actions. This information is easier to use than partial ranking but does reflect less reliable preferences. This information is of interest for assessing the general robustness of the ranking and useful to perform a sensitivity analysis. The Geometrical Analysis for Interactive Aid (GAIA) plane provides the decision maker with a comprehensive graphical image of the decision problem and it is complementary to the multi-criteria analysis. This tool provides clear graphical information regarding the conflicting characters of the criteria and about the weights on the final decision. By examining a GAIA graph it is possible to see where the action lies in relation to the criteria, measure how intensely the criteria affect each action and thereby identify the criteria that are in line with or that conflict with the various alternatives. The decision making process of an energy project is the closing link in the process of analysing and handling different types of information: environmental, technical economic and social. As this study demonstrates, multi-criteria analysis can provide a technical-scientific decision making support tool that is able to justify its choices clearly and consistently.

Nixon et al. (2013) carried out design analysis of concentrated solar thermal collector for industrial application using multi criteria decision making methodology. They developed and evaluated three novel CST concepts derived from the LFR through the following techniques: Quality Function Deployment (QFD), the AHP and the Pugh selection matrix. Criteria were specified by technical and customer requirements gathered from Gujarat, India. The concepts were compared to a standard LFR for reference, and as a result, a novel Elevation Linear Fresnel Reflector (ELFR) concept using elevating mirrors was selected. A detailed version of this concept was proposed and compared against two standard LFR configurations, one using constant and the other using variable horizontal mirror spacing. Annual performance was analysed for a typical meteorological year. Financial assessment was made through the construction of a prototype. The novel LFR had an annual optical efficiency of 49% and increases exergy by 13 to 23%. Operational hours above a target temperature of 300°C were increased by 9% to 24%. A 17% reduction in land usage was also achievable. However, the ELFR suffered from additional complexity and a 16% to 28% increase in capital cost. It was concluded that this novel design is particularly promising for industrial applications and locations with restricted land availability or high land costs. The decision analysis methodology adopted was considered to have a wider potential for applications in the fields of renewable energy and sustainable design.

All these multi criteria decision making methods have validated their effectiveness in a number of solar thermal energy systems. However, such MCDM approaches for implementation in solar thermal collectors lacks in majority which are not only economically efficient but also confirms about effective energy utilization These MCDM methods usually consists of three basic stages viz. identification of the alternatives and criterions, criterion weight calculation and selection of the best alternative. The thermal, hydraulic and thermo-hydraulic criteria of performance evaluation for the solar collectors could thus be compared simultaneously in order to obtain an energy efficient design of solar thermal systems using MCDM methods.

MCDM Techniques for Solar Flat Plate Collector

Rohith et al. (2012) focused on selection of best polymer to manufacture solar flat plate collector, because polymers have its own advantages over conventional metals. Finally, best material was identified by considering different criteria using AHP. Important criteria were selected after defining the criteria and different experts were asked to identify the importance of each criterion by using Saaty's scale. Structuring the hierarchical model involved building the AHP hierarchy model based on the identified criteria, which contained five levels: the goal, the criteria, and alternatives. Formation of pairwise comparison matrix involved the relative importance of the criteria in achieving the goal. The criteria are density, thermal conductivity, cost, safety level temperature, tensile strength, coefficient of thermal expansion, notch impact strength. They were compared with each other in order to determine the relative importance of each factor to accomplish overall criteria. According to expert judgment density is not very strongly preferred than thermal conductivity factor with respect to the best polymer. This was followed by the calculation of Eigen vectors, principal Eigen value, and consistency index and consistency ratio. This may be considered as the best evaluation methodology where vagueness and uncertainty are not involved. High Density Polyethylene (HDPE) was found to be the best material for manufacturing solar flat plate collector.

MCDM Techniques for Solar PV Panel Selection

Various literature reviews suggest that there are a large number of decision making methods and tools available for solar panel selection. Striking a balance between the conflicting tangible and intangible criteria is necessary from a selection point of view. Due to the multi objective nature of the solar panel selection process, there is more work highlighting the use of quantitative models based decision making methods rather than addressing the issue of subjective (qualitative) criteria. A decision model in accommodating both subjective and objective criteria with accurate and faster calculation procedure in making the solar panel selection simpler and user friendly is desirable. This warrants the development of a new decision support system in order to resolve MCDM to the solar panel selection problem. Various factors such as multiple criteria, conflicts amongst criteria, involvement of many alternatives, internal and external constraints imposed on the buying process make the solar panel selection decision making process rather difficult and complicated. The total cost of the solar panel depends on size (in W), the brand, the physical size, the longevity /durability and any certifications of the panel. Selection of a panel is heavily influenced by electrical properties such as power rating, temperature coefficient, peak efficiency, energy density, etc. Some other parameters include its weight, quality and service support. The AHP method for the selection of a suitable solar panel has been found to undergo nearly crisp decision applications. It does not take into account the uncertainty and risk in assessing the alternative's potential performance because it assumes that the relative importance of criteria affecting alternative's performance is certain. The subjective judgment, selection and preference of decision makers result in large influence. However, in a proposed integrated model based on AHP and GRA, both qualitative and quantitative criteria are considered simultaneously in the evaluation and selection of solar panels. The quantitative criteria (cost, peak efficiency, energy density and weight) have the characteristic of "smaller-is-better" as opposed to the "larger-is-better" for the qualitative criteria [service and quality] (Gnanasekaran and Venkatachalama, 2019).

MCDM Techniques for a Solar PV Power Plant

A solar PV power plant installation problem considering both qualitative and quantitative factors was solved in two ways; investigating real data measurements and using AHP method. In this research, first, azimuth angle was found for three locations proposed and then each location at the certain azimuth angle was investigated in terms of several parameters and the best place for installation of a solar PV power plant was evaluated. Second, in order to take several conflicting criteria into account, AHP was applied to that location selection problem proposed considering five main criteria; potential energy, production, environmental factors, safety, distance from existing transmission line and topographical properties. The same result was achieved using AHP method. An examination of different places using both tangible and intangible information was provided. Uncertainties could also arise while consequents of the same problem are defined by different experts, linguistically. Due to uncertainty in subjective judgments of decision makers, the lack of information and conflicts among the criteria, fuzzy logic could be applied to the same problem as a future study. In addition, another possible research direction which could be pursued to extend the work by taking a smaller interval than 5 for azimuth angle range was proposed (Ozdemir & Sahin, 2018).

MCDM Techniques in Other Engineering and Solar Energy Systems

Although MCDM techniques have become popular in the decision making process for performance optimization of engineering systems, the application of the same in modeling and optimization of solar thermal systems are only a few as seen from the last section. However, in general MCDM techniques are finding wide-spread applications in the renewable energy sector including solar energy. Beltran et al. (2014) applied analytical hierarchy process and analytical network process to for the selection of solar-thermal power plant investment projects in Spain. The score-based MCDM technique, called SAW was utilized in the assessment of renewable energy vulnerability in China [40]. SAW method was also utilized for design optimization of Stirling engine, which can efficiently harness radiant flux of Sun to produce solar energy (Wang et al. 2014). An increase of efficiency and reduction of the power loss of Stirling engine were made possible by decision making done through this method (Luo et al. 2016). GRA is another popular MCDM technique, which is being applied in renewable energy sector. It was used to prioritize the energy producing system considering technical, economic, environmental, social and institutional factors (Ebrahimi and Rahmani, 2019), and it was reported that natural gas systems are ranked highest followed by solar and wind energy systems. This method was also utilized to develop the sustainability indicators of renewable energy selected from environmental, economic and social sustainability assessments (Liu et al. 2013). GRA was integrated with principal component analysis method to optimize the performance parameters of a solar driven organic Rankine cycle (Tiwari et al. 2017). Data Envelopment Analysis based Ranking (DEAR) was adopted in analysing the efficiency of common solar stills and Nano-PCM based solar stills (Winfred et al. 2018). It was also used for selection of best material for latent heat energy storage for the application of solar energy based desalination system (Winfred et al. 2019).

MCDM techniques were also adopted in the field of solar PV technologies. DEAR method was utilized to analyze 14 areas of Khuzestan province (Iran), to determine optimum photovoltaic plant location based on four aspects which are the net cost of construction, earnings, the amount of power generated, and pollution produced per plant(Mostafaeipour et al. 2016). Data development analysis was applied for performing a comparative analysis of the crystalline silicon and thin-film photovoltaic solar cells based on four factors i.e. technical factor, economic factor, payback period factor and total factor (Haeri, 2017). Mousazadeh et al. (2009) reviewed different types of sun-tracking systems for solar PV system and their pros and cons. The most efficient and popular sun-tracking device was found to be in the form of polar-axis and azimuth/elevation types by experimental investigation. Chin et al. (2011) proposed the design, modelling and testing of an active single axis solar tracker. The compactness of the proposed solar tracker enables it to be mounted on to the wall. The solar irradiance is detected by two Light-Dependent Resistor (LDR) sensors that are located on the surface of the photo-voltaic (PV) panel. Koussa et al. (2012) investigated the effect of using different types of sun tracking systems on the flat plate solar PV system performance for different sky conditions. And the performance of PV panels mounted on fixed structures inclined according to a yearly and seasonal optimum slope, panels fixed on to single axis, sun tracking system with vertical and inclined rotating axis, for which a yearly and seasonal optimum slopes have also been used, as well as of those mounted on two-axis sun tracker mechanisms have been considered and analysed employing data collected over thirty one selected days. MCDM techniques have proved its performance in converting qualitative measures to quantitative numeric that can help draw a conclusion or make a decision. A knowledge based data for solar tracking system selection considering the response, reliability, repeatability and accuracy is generally desired.

SUMMARY

Solar thermal systems are popular renewable energy systems which harness abundantly available Sun's radiant energy into heat as well as solar thermal power. However, the basic requirement is to have an efficient system so that renewable energy systems can be popularized, which is also required for large-scale integration of such systems. As discussed, solar thermal systems have different input parameters in the form of design, operating, and climatic parameters; similarly there are output parameters like thermal efficiency, outlet fluid temperature, useful heat gain, etc. The optimization process requires an in-depth parametric study of the effect of design and operating parameters on the performance of the solar thermal systems. Similarly optimization of solar thermal system performance shall require consideration of all the operating and design parameters that are conflicting in nature, given the uncertainty of some of the climatic parameters pertaining to the ambient conditions. Hence to address these issues MCDM techniques can be quite useful. However, their benefits to be exploited require thorough analysis of the MCDM models in the performance study of solar thermal systems.

From the study of MCDM techniques in solar thermal systems, it appears that most of the studies were concentrated on heat transfer augmentation of flat plate collectors and concentrated solar thermal collectors. Such studies are also very less. Similarly, the number of studies involving other solar thermal systems like line focussing parabolic trough collector, non-concentrating evacuated tubular collector are hardly available. Thus there is wide scope for implementation of many other individuals plus hybrid MCDM techniques for modeling and optimization of other solar energy system.

REFERENCES

Acır, A., Emin, M., & Cakırog, R. (2017). Parametric optimization of energy and exergy analyses of a novel solar air heater with grey relational analysis. *Applied Thermal Engineering*, *122*, 330–338. doi:10.1016/j.applthermaleng.2017.05.018

Anwar, S., Efstathiadis, H., & Qazi, S. (2013). *Handbook of Research on Solar Energy Systems and Technologies*. doi:10.4018/978-1-4666-1996-8

Aragonés-Beltrán, P., Chaparro-González, F., Pastor-Ferrando, J. P., & Pla-Rubio, A. (2014). An AHP (Analytic Hierarchy Process)/ANP (Analytic Network Process)-based multi-criteria decision approach for the selection of solar-thermal power plant investment projects. *Energy*, *66*, 222–238. doi:10.1016/j.energy.2013.12.016

Avezova, N. R., Rakhimov, E. Y., Khaitmukhamedov, A. E., Boliev, B. B., & Usmonov, A. Y. (2018). Dependence of Techno-Economic and Ecological Indicators of Flat-Plate Solar Water Heating Collectors in Hot Water Supply Systems from the Temperature of Heating Water. *Applied Solar Energy*, *54*(4), 297–301. doi:10.3103/S0003701X18040035

Babaey, B. D., Kharchenko, V., Panchenko, V., & Vasant, P. (2019). Materials and Methods of Thermal Energy Storage in Power Supply Systems. In V. Kharchenko & P. Vasant (Eds.), *Renewable Energy and Power Supply Challenges for Rural Regions* (pp. 115–135). IGI Global.

Bouya, M., El Majd, B. A., Agdid, M., & Abdellah, A. B. (2015). Solar Heater Technologies Selection Using Multi-criteria Analysis. *Journal of Energy and Power Engineering*, *9*(4), 393–398.

Campos-guzmán, V., García-cáscales, M. S., Espinosa, N., & Urbina, A. (2019). Life Cycle Analysis with Multi-Criteria Decision Making : A review of approaches for the sustainability evaluation of renewable energy technologies. *Renewable & Sustainable Energy Reviews*, *104*, 343–366. doi:10.1016/j.rser.2019.01.031

Castro, D.M., & Parreiras, F.S. (2018). A Review on Multi-Criteria Decision-Making for Energy Efficiency in Automotive Engineering. *Applied Computing and Informatics,* 1-13.

Cavallaro, F. (2009). Multi-criteria decision aid to assess concentrated solar thermal technologies. *Renewable Energy*, *34*(7), 1678–1685. doi:10.1016/j.renene.2008.12.034

Cavallaro, F. (2010). Fuzzy TOPSIS approach for assessing thermal-energy storage in concentrated solar power (CSP) systems. *Applied Energy*, *87*(2), 496–503. doi:10.1016/j.apenergy.2009.07.009

Chamoli, S. (2015). Hybrid FAHP (fuzzy analytical hierarchy process)-FTOPSIS (fuzzy technique for order preference by similarity of an ideal solution) approach for performance evaluation of the V down perforated baffle roughened rectangular channel. *Energy*, *84*, 432–442. doi:10.1016/j.energy.2015.03.007

Chamoli, S. (2015). Preference selection index approach for optimization of V down perforated baffled roughened rectangular channel. *Energy*, *93*, 1418–1425. doi:10.1016/j.energy.2015.09.125

Chauhan, R., Singh, T., Thakur, N. S., & Patnaik, A. (2016). Optimization of parameters in solar thermal collector provided with impinging air jets based upon preference selection index method. *Renewable Energy*, *99*, 118–126. doi:10.1016/j.renene.2016.06.046

Chin, C. S., Babu, A., & McBride, W. (2011). Design, modeling and testing of a standalone single axis active solar tracker using MATLAB/Simulink. *Renewable Energy*, *36*(11), 3075–3090. doi:10.1016/j.renene.2011.03.026

Diemuodeke, E. O., Addo, A., Oko, C. O. C., Mulugetta, Y., & Ojapah, M. M. (2018). Optimal Mapping of Hybrid Renewable Energy Systems for Locations Using Multi-Criteria Decision-Making Algorithm. *Renewable Energy*, *•••*, 1–6.

Duffie, J. A., & Beckman, W. A. (1980). *Solar thermal engineering processes* (4th ed.). Wiley.

Ebrahimi, M., & Rahmani, D. (2019). A five-dimensional approach to sustainability for prioritizing energy production systems using a revised GRA method : A case study. *Renewable Energy*, *135*, 345–354. doi:10.1016/j.renene.2018.12.008

Gal, T., Stewart, T., & Hanne, T. (Eds.). (2013). *Multi criteria decision making: advances in MCDM models, algorithms, theory, and applications* (Vol. 21). Springer.

Gnanasekaran, S., & Venkatachalama, N. (2019). A review on applications of multi-criteria decision making (MCDM) for solar panel selection. *International Journal of Mechanical and Production Engineering Research and Development*, *9*(2), 11–20. doi:10.24247/ijmperdapr20192

Groenendaal, B. J. (2002). *Solar thermal power technologies, Monograph in the framework of the VLEEM Project.* Energy Research Centre of the Netherlands.

Haeri, A. (2017). Evaluation and comparison of crystalline silicon and thin-film photovoltaic solar cells technologies using data envelopment analysis. *Journal of Materials Science Materials in Electronics*, *28*(23), 18183–18192. doi:10.100710854-017-7764-y

Hwang, C. L., & Yoon, K. (2012). *Multiple attribute decision making: methods and applications a state-of-the-art survey* (Vol. 186). Springer.

Jackson, C., & Pascual, R. (2008). Optimal maintenance service contract negotiation with aging equipment. *European Journal of Operational Research*, *189*(2), 387–398. doi:10.1016/j.ejor.2007.05.039

Jeong, B., Oguz, E., Wang, H., & Zhou, P. (2018). Multi-criteria decision-making for marine propulsion : Hybrid, diesel electric and diesel mechanical systems from cost-environment-risk perspectives. *Applied Energy*, *230*, 1–17. doi:10.1016/j.apenergy.2018.09.074

Kalogirou, S. A. (2004). Solar thermal collectors and applications. *Progress in Energy and Combustion Science*, *30*(3), 231–295. doi:10.1016/j.pecs.2004.02.001

Kharchenko, V., & Vasant, P. (2018). *Handbook of Research on Renewable Energy and Electric Resources for Sustainable Rural Development.* doi:10.4018/978-1-5225-3867-7

Kharchenko, V., & Vasant, P. (2019a). Handbook of Research on Energy-Saving Technologies for Environmentally-Friendly Agricultural Development. Academic Press.

Kharchenko, V., & Vasant, P. (2019b). Handbook of Research on Smart Computing for Renewable Energy and Agro-Engineering. Academic Press.

Kharchenko, V., & Vasant, P. (2019c). *Renewable Energy and Power Supply Challenges for Rural Regions.* Academic Press.

Klychev, S. I., Bakhramov, S. A., & Harchenko, V. V. (2018). Non stationarity of the Efficiency Coefficient and Heating Temperatures of Water in Flat-Plate Solar Collectors. *Applied Solar Energy*, *54*(4), 287–292. doi:10.3103/S0003701X18040072

Koussa, M., Haddadi, M., Saheb, D., Malek, A., & Hadji, S. (2012). Sun tracker systems effects on flat plate photovoltaic PV systems performance for different sky states: A case of an arid and hot climate. *Energy Procedia, 18*, 839–850. doi:10.1016/j.egypro.2012.05.098

Kreith, F., & Kreider, J. F. (1978). *Principles of solar engineering*. McGraw-Hill.

Kuo, C.J., Su, T., Jhang, P., Huang. C., & Chiu, C. (2011). *Using the Taguchi method and grey relational analysis to optimize the flat-plate collector process with multiple quality characteristics in solar energy collector manufacturing*. Academic Press.

Liu, G., Baniyounes, A. M., Rasul, M. G., Amanullah, M. T. O., & Khan, M. M. K. (2013). General sustainability indicator of renewable energy system based on grey relational analysis. *International Journal of Energy Research, 37*(14), 1928–1936. doi:10.1002/er.3016

Luo, Z., Sultan, U., Ni, M., Peng, H., Shi, B., & Xiao, G. (2016). Multi-objective optimization for GPU3 Stirling engine by combining multi-objective algorithms. *Renewable Energy, 94*, 114–125. doi:10.1016/j.renene.2016.03.008

Mardani, A., Jusoh, A., Nor, K., Khalifah, Z., Zakwan, N., & Valipour, A. (2015). Multiple criteria decision-making techniques and their applications–a review of the literature from 2000 to 2014. *Economic Research-. Ekonomska Istrazivanja, 28*(1), 516–571. doi:10.1080/1331677X.2015.1075139

Mostafaeipour, A., Qolipour, M., & Mohammadi, K. (2016). Evaluation of installing photovoltaic plants using a hybrid approach for Khuzestan province, Iran. *Renewable & Sustainable Energy Reviews, 60*, 60–74. doi:10.1016/j.rser.2016.01.105

Mousazadeh, H., Keyhani, A., Javadi, A., Mobli, H., Abrinia, K., & Sharifi, A. (2009). A review of principle and sun-tracking methods for maximizing solar systems output. *Renewable & Sustainable Energy Reviews, 13*(8), 1800–1818. doi:10.1016/j.rser.2009.01.022

Nixon, J. D., Dey, P. K., & Davies, P. A. (2013). Design of a novel solar thermal collector using a multi-criteria decision-making methodology. *Journal of Cleaner Production, 59*, 150–159. doi:10.1016/j.jclepro.2013.06.027

Ozdemir, S., & Sahin, G. (2018). Multi-criteria decision-making in the location selection for a solar PV power plant using AHP. *Measurement, 129*, 218–226. doi:10.1016/j.measurement.2018.07.020

Rohith, B. J., Venkataramaiah, P., & Mohana Reddy, P. (2012). Material selection for solar flat plate colletors using AHP. *International Journal of Engineering Research and Applications*, *2*(2), 1181–1185.

Sabaei, D., Erkoyuncu, J., & Roy, R. (2015). A review of multi-criteria decision making methods for enhanced maintenance delivery. *Procedia CIRP*, *37*, 30–35. doi:10.1016/j.procir.2015.08.086

Sharma, A., Chauhan, R., Singh, T., Kumar, A., Kumar, R., & Sethi, M. (2017). Optimizing discrete V obstacle parameters using a novel Entropy-VIKOR approach in a solar air flow channel. *Renewable Energy*, *106*, 310–320. doi:10.1016/j.renene.2017.01.010

Sharma, A., Chauhan, R., Singh, T., Kumar, A., Kumar, R., & Sethi, M. (2017). Optimizing discrete V obstacle parameters using a novel Entropy-VIKOR approach in a solar air flow channel. *Renewable Energy*, *106*, 310–320. doi:10.1016/j.renene.2017.01.010

Sukhatme, S. P., & Nayak, J. K. (2017). *Solar energy*. McGraw-Hill Education.

Sychov, A., Kharchenko, V., Vasant, P., & Uzakov, G. (2018). Application of Various Computer Tools for the Optimization of the Heat Pump Heating Systems with Extraction of Low-Grade Heat from Surface Watercourses. In *International Conference on Intelligent Computing & Optimization* (pp. 310-319). Springer.

Tiwari, D., Sherwani, A.F., Asjad, M., & Arora, A. (2017). Grey relational analysis coupled with principal component analysis for optimization of the cyclic parameters of solar driven organic rankine cycle. *Gray Systems: Theory and Applications, 7*(2), 218-235.

Vladislav, P., & Martin, L. (2010). Photovoltaics. Solar Energy.

Wang, B., Ke, R., Yuan, X., & Wei, Y. (2014). China's regional assessment of renewable energy vulnerability to climate change. *Renewable & Sustainable Energy Reviews*, *40*, 185–195. doi:10.1016/j.rser.2014.07.154

Winfred, D. D., Kumar, V. R., Suganthi, L., Iniyan, S., & Davies, P. A. (2018). Techno-economic analysis of solar stills using integrated fuzzy analytical hierarchy process and data envelopment analysis. *Solar Energy*, *159*, 820–833. doi:10.1016/j.solener.2017.11.050

Winfred, D. D., Rajkumar, V., Suganthi, L., & Iniyan, S. (2019). Solar Energy Materials and Solar Cells Studies on latent heat energy storage (LHES) materials for solar desalination application-focus on material properties, prioritization, selection and future research potential. *Solar Energy Materials and Solar Cells*, *189*, 149–165. doi:10.1016/j.solmat.2018.09.031

Chapter 2
Modeling and Optimization of Solar Flat Plate Collector

ABSTRACT

Solar flat plate collector (SFPC) is a heat exchanger that transforms radiant solar energy into thermal energy in the form of heated fluid. The performance of SFPC is very much dependent on operating/input and response/output parameter which mainly affects the efficiency of SFPC. This chapter presented the modeling and optimization of SFPC system parameters (solar radiation [I], wind velocity [V], ambient temperature [Ta], and Inlet Temperature [Ti]) for SFPC. Modified-fuzzy set theory with MOOSRA (M-FST-MOOSRA) was employed to optimize the SFPC system. Based on results, trail no. 14 (i.e., I = 825 W/m2, V = 1.4 m/s, Ta = 28.8oC, and Ti = 66.4oC) gave highest RPI among the other trail nos. and shows the optimal setting which results in higher efficiency and better performance for the SFPC. Further, parametric analysis is also done to determine the most important parameter followed by analysis of variance (ANOVA) analysis. Last, confirmatory test are conducted to verify and validate the proposed method with the experimental results.

INTRODUCTION

Solar flat plate collector is a heat exchanger to transform solar energy into thermal (heat) energy that is used for various domestic and industrial applications, for e.g. domestic water heating, space heating, drying, distillation, etc. Solar water heating systems can be an active system mounted on the roof

DOI: 10.4018/978-1-7998-3523-3.ch002

top of buildings. A photograph (Courtesy- Ministry of New and Renewable Energy, MNRE, India) of such roof top systems is given in Figure 1a. For a solar water heating system (Figure 1b), the major components are a tilted flat plate collector (at an angle equal to the latitude of the place), an insulated hot water storage tank, and the pipelines. For a solar air heating system, the major components are a flat plate collector, a centrifugal blower, and a dryer cabin. In case of water heating, the purpose of the insulated tank is to store hot water coming out of the collector. The tank can store hot water for 2-3 days with very less temperature drop, normally of the order of 5°-7°C. The solar radiation is absorbed in the flat plate collector that consists of a rectangular metallic collector box covered on the top with a toughened glass sheet and insulated on its bottom by layers of puff insulation. Inside there is a blackened metallic absorber (selectively coated) sheet made from Al/Cu with built in channels or riser tubes made from copper to carry water (Figure 2). The absorber plate absorbs the solar radiation and transfers the heat to the water flowing inside the riser tube. The insulated storage tank is usually located slightly above the collector. As solar energy heats up the water in the riser tubes, water automatically flows to the top of that tank and the empty space is replaced by the hot water. When the hot water is withdrawn from the storage tank for further applications (Ummadisingu & Soni, 2011), cold water automatically enters the storage tank through an opening at the bottom of the storage tank. The cold water is supplied to the storage tank from a separate cold-water tank, which is mounted at least two to three feet above the top of the tilted collector. The hot water storage tank design is shown in Figure 3. The implication of the storage tank for effective performance of the system is immense because it prevents heat loss up to a maximum of 3°-5°C per day, which is due to the thick insulation layer (of about 10 cm thickness) around the storage tank as shown in Figure3. There is an air vent on top of the tank, which is used to let any steam, if formed, to go off the tank. The hot water is stored in the storage tank due to thermal stratification, which means storing hot water in layers having the hottest layers on top of the tank and the coolest on the bottom. The thermo siphon effect is responsible for the thermal stratification to take place. If hot water stored is mixed in the tank then the thermal stratification is destroyed, which is done if uniform temperature of water is required in the storage tank.

On the other hand, performance of SFPC depends on its operating and performance parameters. Improper selection of these parameters results in poor performance in the SFPC performance. Hence, optimization of SFPC parameters is essential. Although many investigators have worked

Figure 1. (a) Roof-top solar water heating system (b) Solar water heating system

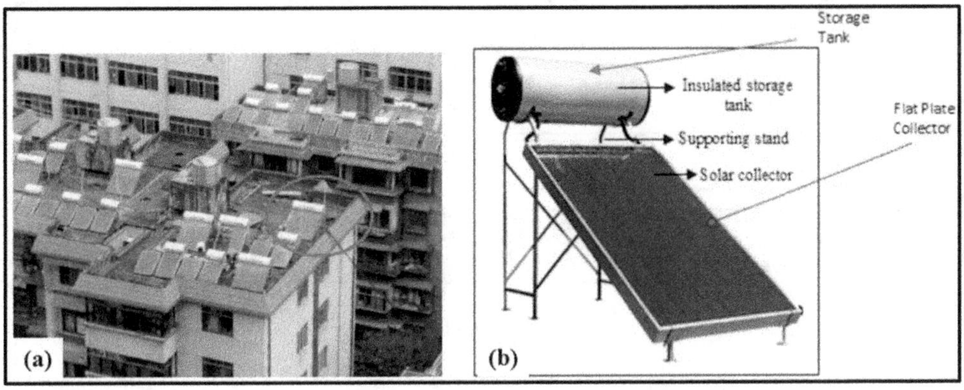

on optimization of solar system (Suresh et. al., 2006; Lahu et al., 2014; Athanasios et al., 2016; Dorfeshan et al.,2018; Bhowmik, 2014) they are limited to single objective optimization. In addition, importance of criteria weights and effect of both beneficial and non-beneficial parameters on the performance of SFPC have been ignored. In order to overcome this, present chapter proposed a hybrid integrated MCDM method which consists of fuzzy set theory (Chamoli, 2015a) with MOOSRA (Chamoli, 2015b; Sharma et al., 2017; Acır, et al., 2017) methods. Here, fuzzy set theory is used for extraction of precise priority weights while MOOSRA for optimal selection of SFPC parameters. This integrated method offers many advantages like flexibility to solve multi-criteria problem, considers multiple criteria with different data measurements, allows degree of uncertainty, imprecise with subjective and objective data during optimization, easy to maintain consistency, less computational time required, very simple and stable technique, minimum mathematical calculation involved. These advantages may be quite useful and helpful to the decision makers who may not have a strong background in mathematics. Also, none of the literature on modeling and optimization of SFPC parameters using Fuzzy-MOOSRA method is available.

Experimental Details

The SFPC's are tested for evaluating their performance following various International Standards. In Indian context, such collectors are tested following Indian Standard, IS 12933 under standard conditions. According to Bureau

Figure 2. Pictorial views of a flat-plate collector including components

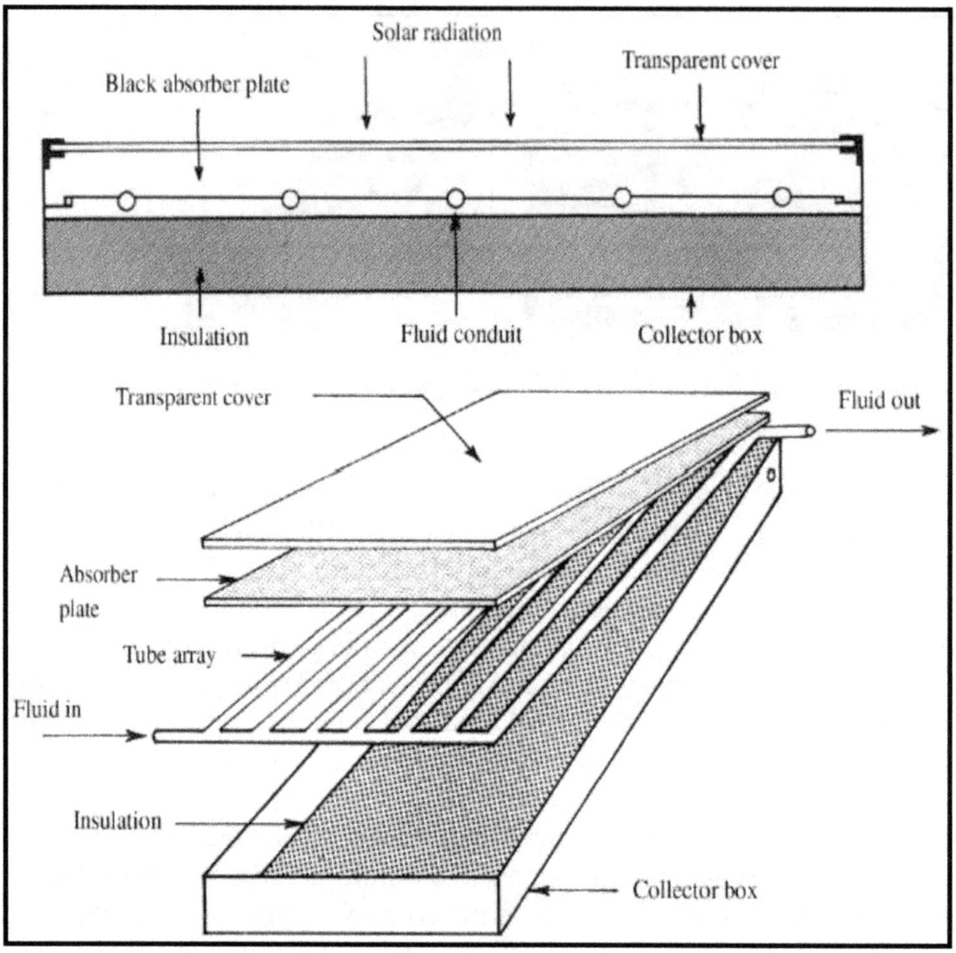

of Indian Standards (IS 12933 Part 5), the test conditions followed during determination of the thermal efficiency are as follows (Kuo, et al., 2011):

a. The total solar irradiance incident on the flat plate collector must not be less than 700 W/m^2. No two consecutive solar radiation values should differ by ±50 W/m^2.
b. The average value of the surrounding air speed must lie between 1-5 m/s.
c. The mass flow rate of the water needs to be fixed at 0.02 kg/s m^2 based on the gross area of the collector.

Figure 3. Detailed design of storage tank

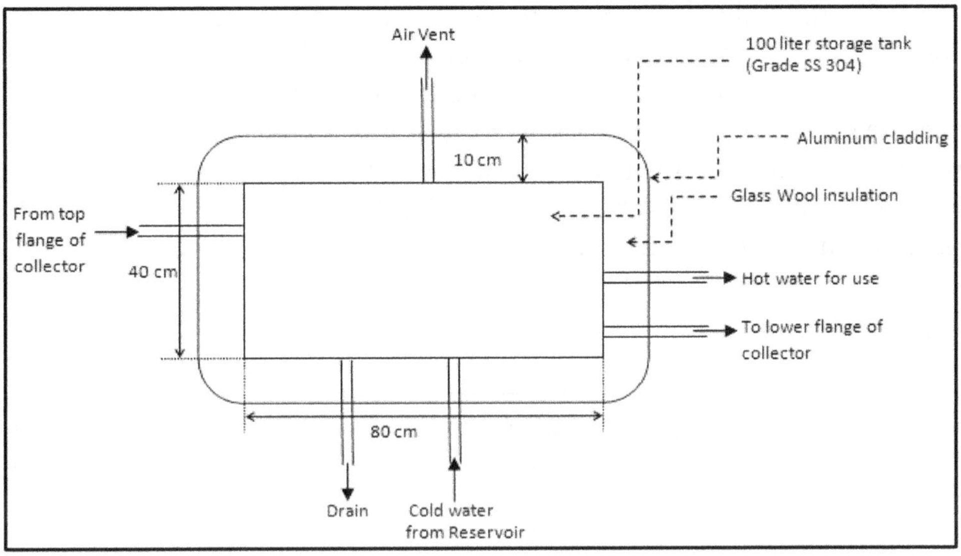

d. The flow rate must be held stable to within ±1% of the set value during each test period and must not vary by more than ±1% of the set value from one test to the other.

The test set up for analyzing the thermal performance of flat plate collector system for water heating consists of a rotameter (range 0-10lph, least count ±1 litre per minute), electromagnetic flowmeter (range 0-5 lpm), constant head tank, geyser, and piping system. A schematic diagram of the test set up is shown in Figure 4a and the photograph of the experimental setup is shown in Figure 4b. The technical details of the experimental setup can be found in Figure 4b, which is also discussed here. It is a closed loop unit where hot water at the set point temperature (Proportionate-Integral-Derivative, PID controlled) is allowed to enter the inlet of SFPC through an electromagnetic flowmeter. The Flat Plate Collector (FPC) is 2 m in length and 1m in breadth. The water coming out from FPC at the outlet is transferred to the storage tank, completing the loop. Water is initially pre-heated in the geyser before entering the constant head tank. There is a centrifugal pump (of fractional horsepower capacity i.e. 0.5hp) at the lower level that lifts water to the level of geyser. Temperature sensors (PT 1000 RTD type with a range 0 to 150°C with an accuracy ± 0.5°C) at inlet and outlet of the FPC are used to measure water inlet and outlet temperature. An ambient temperature sensor (PT 100

RTD type with a range -40° to 60°C with an accuracy ± 0.2°C) is used for measuring the ambient temperature. For measuring flow rate, an electronic flow meter is used (range 0.5-25 lpm, accuracy ± 1%). A solar pyranometer is mounted on the plane of the tilted collector (as shown in Figure 4b) to measure global solar radiation (range 0-1500 W/m^2 with an accuracy ± 1 W/m^2). All the readings of temperature and intensity are recorded in a 16-channel datalogger (range 50 m/s, 60°C, 40mV, 200 °C with an accuracy of ± 1%) for retrieving the required data later on.

Figure 4a. Line diagram of the test set up of SFPC

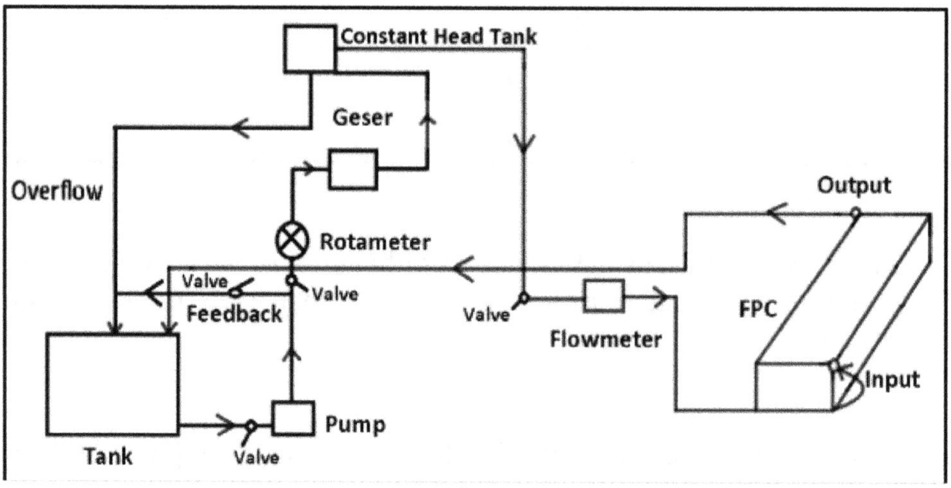

The major performance parameters of solar flat plate collector based water heater are its thermal efficiency (η), exergetic efficiency (η_{Ex}), and useful heat gain (*h*), which depend on ambient and operating parameters such as intensity of solar radiation (*I*), ambient temperature (T_a), water temperature at inlet to the collector (T_i), outlet temperature from the collector (T_o). In addition to these ambient and operating parameters, performance also depends on the collector tilt angle and height gap between the hot water storage tank and the collector. The thermal efficiency can be expressed using the following equation-

$$\eta = F_R\,(\tau\alpha)e\,_{-}\,(FR\,_{U}L)\,x\,(Ti\,\_Ta)_{/}I \quad (1)$$

Figure 4b. Experimental setup showing flat plate collector (absorber area 2 m2) and constant pressure head tank for supply of hot water

where first term is the product of heat removal factor (F_R) and effective absorptivity-transmittivity product, $(\tau\alpha)_e$, and the second term is the product of heat removal factor and heat loss coefficient (UL). Exergetic efficiency can be calculated in the following manner-

$$\text{Exergy of the Sun, } Ex_S = IA\left(1-\frac{T_a}{T_S}\right) \tag{2}$$

Exergy rise of water can be calculated using the following equation-

$$Ex_W = mc_p\left(T_O - T_i\right) - mc_pT_a ln\left(\frac{T_o}{T_i}\right) \tag{3}$$

where T_S is the temperature of the Sun (5762 K), A is the surface area of the collector, c_p is the specific heat of water, which is 4.18 kJ/kg-K. Therefore, the exergetic efficiency of the solar flat plate collector is calculated using the following equation-

$$\eta_{Ex} = \frac{Ex_W}{Ex_S} \tag{4}$$

Table 1. Experimental Results of SFPC under standard experimental conditions

Trail. no.	Input Parameters				Output Parameters			
	Irradiation (I) in W/m²	Wind velocity(V) in m/s	Ambient Temperature (T_a) in °C	Inlet Temperature (T_i) in °C	Outlet Temperature (T_o) in °C	Efficiency (η) in %	Useful heat gain (h) in W	Exergetic Efficiency (η_{Ex}) in %
1	769	1.4	32.7	76.1	69.9	67.50	519.064	8.421
2	763	1.6	32.7	76.1	69.8	69.13	527.436	8.614
3	760	1.2	32.6	76.2	70.1	67.20	510.692	8.432
4	740	1.3	32.7	76.1	70.0	69.01	510.692	8.619
5	806	2.1	23.8	47.0	40.8	64.40	519.064	4.364
6	797	2.0	23.8	47.3	40.7	69.33	552.552	4.721
7	801	1.9	23.7	47.4	40.6	71.07	569.296	4.862
8	807	1.7	23.8	47.5	40.7	70.54	569.296	4.825
9	841	1.8	28.1	57.0	50.5	64.71	544.180	5.433
10	841	1.9	28.2	57.0	50.4	65.70	552.552	5.486
11	844	1.6	28.2	57.0	50.4	65.47	552.552	5.467
12	847	1.7	28.5	57.1	50.3	67.21	569.296	5.546
13	820	1.5	28.8	66.4	60.5	60.24	493.948	6.638
14	825	1.4	28.8	66.4	60.6	58.86	485.576	6.495
15	813	1.1	28.8	66.5	60.6	60.76	493.948	6.713
16	798	1.3	28.9	66.5	60.7	60.85	485.576	6.712

The useful heat gain by the collector is equal to the enthalpy gain due to heating of water in the collector, which can be calculated using the following equation-

$$Q_{useful} = mc_p(T_O - T_i) \tag{5}$$

The data generated by following the standard test conditions, as described above, are given in Table 1 and using Eqs. (1)- (5), the thermal efficiency (η) and the other performance parameters are calculated. These data are generated for a standard mass flow rate of 2.4 *lpm*. There are different sensors measuring the solar radiation, ambient temperature, fluid temperature, and flow rate. All the data can be logged in a Datalogging facility available in solar test center. In the present case, the necessary facilities of the Regional test center of National Institute of Technology Silchar, Assam, India have been utilized for experimentations on the SFPC. The accuracies of all the measuring instruments are given in Table 2.

Table 2. Details of measuring instruments

Measuring instruments	Specifications	Accuracy
Pyranometer	0-4000 W/m^2	±8 W/m^2
Temperature Sensor (RTD PT1000)	0-150°C	±1.6°C
Rotameter	0-5 LPM	1.196LPM
Data logger	0-200°C & 0-40mV DC	Expanded uncertainty at 95% confidence level is ± 0.1 °C & 0.001mV DC

Integrated M-FST-MOOSRA Method

An Integrated MCDM method (i.e. M-FST-MOOSRA) is presented for modeling and optimization of SFPC system. The hybrid method consists of modified fuzzy set theory (Chamoli, 2015a) with MOOSRA (Chamoli, 2015b; Sharma et al., 2017; Acır, et al., 2017). In this method, extraction of criteria weights is done via fuzzy set theory which influences the response of SFPC while MOOSRA method for optimization of SFPC parameters considering the effect of both beneficial and non-beneficial parameter into account. In the earlier experimental work on parametric investigations for modeling the collector performance, there are a number of operating conditions to be controlled that influence the collector performance, for e.g. higher solar radiation, wind speed and ambient temperature directly influence thermal losses from the collector. These again necessitate entailing conflicting design requirements of the collector, for e.g. absorber area. For various decision variables (absorber area, tilt angle etc.) from the experimental setup that control the response of the solar thermal collectors, a decision making tool like MCDM for initial design selection is a necessity. The decision variables related with the design and operating parameters of the experimental setup, also called criterias, have different weightage of their influences on the collector performance. As under variable conditions, these decision variables cannot be accurately predicted, hence fuzzy based theories are used to extract the criteria weights. The output of the analysis done using fuzzy set theory is then integrated with an MCDM decision making tools to complete the final task, i.e. optimize the collector design and operating parameters for improved performance. The combination of fuzzy theory with MCDM technique is then called hybrid MCDM technique. The schematic diagram of the proposed method is shown in Figure 5.

The detail steps of the hybrid method are as follows:

Figure 5. Schematic flow chart of integrated M-FST-MOOSRA method

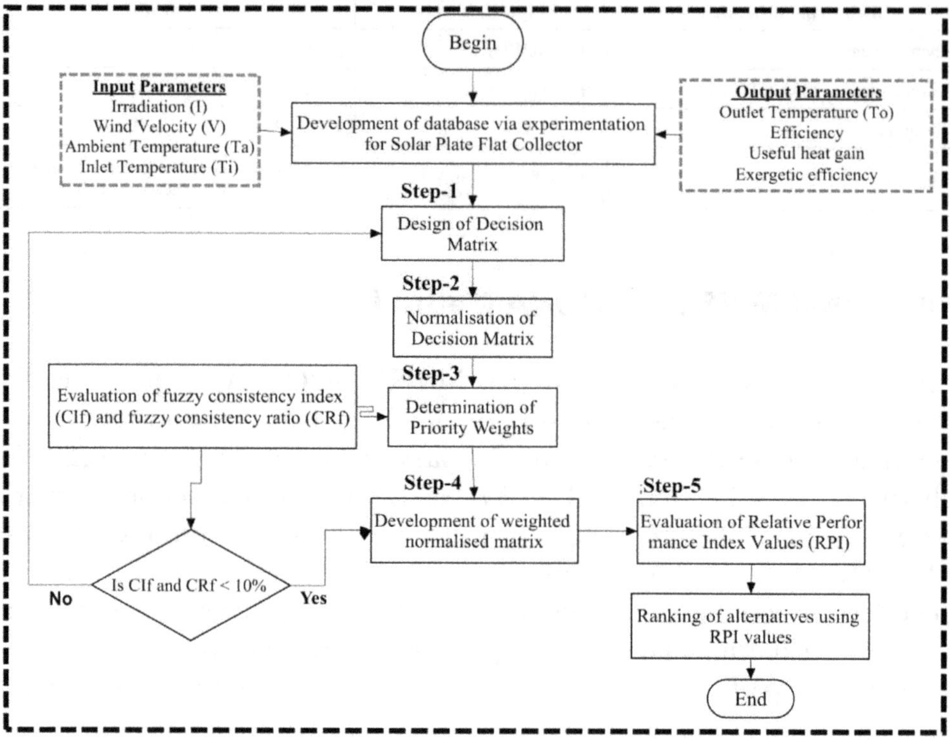

Step1: Design of decision matrix

The SFPC optimization starts with the design of decision matrix. The decision matrix includes number of operating variables i.e. input parameters and number of response variables i.e. output parameters. The typical decision matrix is designed using following equation (Chamoli, 2015b).

$$P_{ij} = \begin{bmatrix} P_{11} & P_{12} & \cdots\cdots & P_{1n} \\ P_{21} & P_{22} & \cdots\cdots & P_{2n} \\ \vdots & \vdots & \vdots & \vdots \\ \vdots & \vdots & \vdots & \vdots \\ P_{m1} & P_{m2} & \cdots\cdots & P_{mn} \end{bmatrix} \text{For } (i=1,\ 2,\ \ldots\ldots m;\ j=1,2,\ \ldots\ldots n) \quad (6)$$

where P_{11}, P_{12}, P_{13}..... P_{mn} are performance values of *j*th criteria to *i*th alternative, P_{ij} denotes decision matrix, *n* is the number of criteria, and *m* denotes the number of alternatives.

Step2: Normalization of the decision matrix

Normalization of the decision matrix is done to get the normalized values of criterias i.e. conversion of different units of measurements into comparable units in the range of 0-1. This process is essential because every process contains different measurement data that result in difficulties in comparison and quantification of different criteria into single unit during the analysis (Chamoli, 2015b; Sharma et al., 2017). The normalized values indicate the relative performance of the generated design criteria/output parameters. The normalization of the different operating variables is done using the following expressions:

For beneficial criteria

$$\left[N_{ij}\right] = \frac{P_{ij}}{\sum_{i=1}^{k} P_{ij}} \text{ For } i=1,2,\ldots,n \text{ and } j=1,2,\ldots,m \tag{7}$$

For Non-beneficial criteria

$$\left[N_{ij}^{*}\right] = \frac{\frac{1}{P_{ij}}}{\sum_{i-k+1}^{k} \frac{1}{P_{ij}}} \text{ For } i=1,2,\ldots,n \text{ and } j=1,2,\ldots,m \tag{8}$$

where *k* the number of beneficial variables, N_{ij} denotes the normalized matrix for beneficial criteria, N_{ij}^{*} denotes the normalized matrix for non-beneficial criteria,

Step 3: Determination of priority weights

In this step, priority weights for each of the criteria are calculated using modified fuzzy theory (Chamoli, 2015a). Here, fuzzy decision matrix is first formulated taking the opinion of the Decision Makers (DM). Each of the DM opinion is then tabulated in terms of linguistic variables using Triangular Fuzzy Number (TFN) as tabulated in Table 3.

Table 3. Linguistic variables (Chamoli, 2015a)

Importance	Fuzzy weight
Extremely low (EL)	(0, 0, 0.1)
Very low (VL)	(0, 0.1, 0.3)
Low (L)	(0.1, 0.3, 0.5)
Medium (M)	(0.3, 0.5, 0.7)
High (H)	(0.5, 0.7, 0.9)
Very high (VH)	(0.7, 0.9, 1)
Extremely high (EH)	(0.9, 1, 1)

Then, aggregation of TFN is done to get the priority weights (ξ_i) for the criteria. Subsequently, consistency of the fuzzy decision matrix and accuracy of the priority weights obtained using fuzzy set theory are verified by fuzzy Consistency Index (CI_f) and fuzzy Consistency Ratio (CR_f) by using the following expressions.

$$CI_f = \frac{(f_{\max} - n)}{n-1} \tag{9}$$

$$CR_f = \frac{CI_f}{RI} \tag{10}$$

where, f_{max} denotes the largest eigenvalue of fuzzy decision matrix, n represents size of comparison matrix or number of criteria's, RI is the random index and its value depends on number of criteria and it is in the range between 0-1.49. The value of CI_f and CR_f should be less than 10%.

Step 4: Development of weighted normalized decision matrix

In this step, formulation of weighted normalized matrix is done using the Eqs. (7-8). The weighted normalization matrix is obtained by multiplying the priority weights of each criteria obtained using fuzzy set theory with each normalized performance value (Sharma et al., 2017; Acır, et al., 2017) of normalized decision matrix.

$$w = \xi i \times [Ni_{jl} \text{ For } i=1,2,\ldots,n \text{ and } j=1,2,\ldots,m \tag{11}$$

$$w^* = \xi_i \times \left[N_{ij}^* \right] \text{For } i=1,2,\ldots,n \text{ and } j=1,2,\ldots,m \tag{12}$$

where, w, w^* represents the weighted normalized decision matrix for both beneficial and non-beneficial criterias, ξ_i is the priority weights of each criteria.

Step5: Relative Performance Index values and ranking

Next, Relative Performance Index (*RPI*) for each of the response parameters is evaluated in order to assess the optimal selection of criteria. This converts the multi-attribute problem into single attribute problem subjected to certain constraints with consideration of both beneficial and non-beneficial criteria into account by using the following expression (Chamoli, 2015b; Acır, et al., 2017).

$$RPI_i = \frac{\sum_{j=1}^{g} w}{\sum_{j=g+1}^{n} w^*} \tag{13}$$

where g represents the number of criteria to be maximized, $(n\text{-}g)$ represents the number of criteria to be minimized; RPI_i represents the relative performance index of i*th* alternatives with respect to all j*th* criterias. At last, ranking of alternatives is done based on the values of *RPI*. The optimal trail no. or parameter setting is selected based on the maximum *RPI* value.

Modeling of SFPC

This section presents the modeling steps of SFPC system using integrated MCDM method. The integrated MCDM method consists of modified fuzzy set theory (Chamoli, 2015a) coupled with MOOSRA method. In this method, modified fuzzy set theory is utilized for determination of precise priority weights of the SFPC parameters while MOOSRA method (Chamoli, 2015b; Sharma et al., 2017) for optimal process parameters of SFPC. First, design of decision matrix is done using Eq. (2), which contains number of criteria as output parameters while number of alternatives as trail no. or input parameter settings. For this, present work considers four input parameters like solar irradiation (*I*), wind velocity (*V*), ambient temperature (T_a), and inlet water temperature (T_i) and three response parameter such as outlet temperature (T_o), efficiency (η), and useful heat gain (*h*) as output parameters. Simultaneously, 16 nos. of experiments are performed on SFPC and corresponding output parameters are determined. The experimental results as tabulated in Table 1 are considered as decision matrix in the present study. Second, normalization of the experimental data is carried out using the Eqs. (3-4). This converts the original sequence data of outlet temperature (T_o), efficiency (η) and useful heat gain(*h*) into comparable sequence data in the range of 0 to 1 and the normalized results are given in Table 4.

Table 4. Normalized values of SFPC parameters

Trail. no.	Normalized values (N_{ij})		
	Outlet Temperature (T_o)	Efficiency (η)	Useful heat gain (h)
1	0.0788	0.0642	0.0614
2	0.0787	0.0657	0.0624
3	0.0791	0.0639	0.0604
4	0.0790	0.0656	0.0604
5	0.0460	0.0612	0.0614
6	0.0459	0.0659	0.0653
7	0.0458	0.0676	0.0673
8	0.0459	0.0671	0.0673
9	0.0570	0.0615	0.0644
10	0.0568	0.0625	0.0653
11	0.0568	0.0622	0.0653
12	0.0567	0.0639	0.0673
13	0.0682	0.0573	0.0584
14	0.0684	0.0560	0.0574
15	0.0684	0.0578	0.0584
16	0.0685	0.0578	0.0574

In the third step, determination of priority weights for each of the SFPC parameters such as outlet temperature (T_o), efficiency (η) and useful heat gain (h) is done using Fuzzy set theory. Fuzzy decision matrix is developed by taking the opinion of the DM. Opinion of four DM is taken in terms of linguistic variable using TFN as tabulated in Table 3 (Chamoli, 2015a). After that, aggregation of TFN for each of the output parameter is done to get the priority weights (ξ_i) and results are tabulated in Table 5. It is observed that the parameter outlet temperature (T_o) yields highest priority weight (ξ_i = 0.7510), which is of more significance than the efficiency (η) and useful heat gain (h) for the SFPC performance. Simultaneously, consistency of the fuzzy decision matrix and accuracy of the priority weights (ξ_i) for each of the SFPC parameters obtained using fuzzy set theory are verified by fuzzy CI_f and fuzzy CR_f using Eqs. (5-6). It is seen that the value of CI_f and $CR_{f\,as}$ obtained as 0.0023 and 0.0036, being less than 10%, results better consistency and accuracy of priority weights (ξ_i).

Table 5. Fuzzy decision matrix & priority weights of SFPC output parameters

Output response	Triangular fuzzy weights			Fuzzy weights(ξ_i)
To	0.4534	0.8564	0.9423	0.7510
η	0.3251	0.5437	0.8653	0.5782
h	0.2562	0.4845	0.7958	0.5122
CI_f = 0.0023 < 10%; $CR_{f=}$ 0.0036 < 10%				

Next, development of weighted normalized decision matrix is obtained using Eqs. (7-8). Here, priority weights (ξ_i) of outlet temperature (T_o) efficiency(η) and useful heat gain (h) are multiplied with each performance value of normalized decision matrix. After the weighted normalized decision matrix, determination of relative performance index values for each of the alternatives is done using Eq. (9). This converts the multi-attribute problem into single attribute problem with considering the effect of both beneficial and non-beneficial parameter and interrelationship between each criteria/ output parameter into account (Sharma et al., 2017; Acır, et al., 2017). At last, ranking of alternatives is obtained based on the *RPI* values; an alternative with the highest *RPI* yields the optimal parameter setting for SFPC system, and results of the same are tabulated in Table 8.

RESULTS AND DISCUSSION

Based on the detailed modeling of the SFPC (thermal modeling and the MCDM modeling), parametric optimization has been done for improving the performance of the same. A detailed parametric analysis is done first using an experimental approach to elucidate how the different input parameters affect the output parameters as also given in Table 1.

Parametric Analysis

The performances of the SFPC are evaluated by analysing the variations of useful heat gain, thermal efficiency (η), and exergetic efficiency (η_{Ex}) over time as well as the variations of these efficiencies with respect to reduced temperature difference. The reduced temperature difference measures the difference of collector inlet temperature and the ambient temperature relative the intensity of solar radiation, i.e. it combines the effect of the increase of inlet temperature to the collector, ambient temperature and the temperature of heat addition from the incident solar radiation on the efficiencies of the flat plate collector. Thus the reduced temperature difference is an important

Figure 6. Plot of solar irradiation and ambient temperature with time

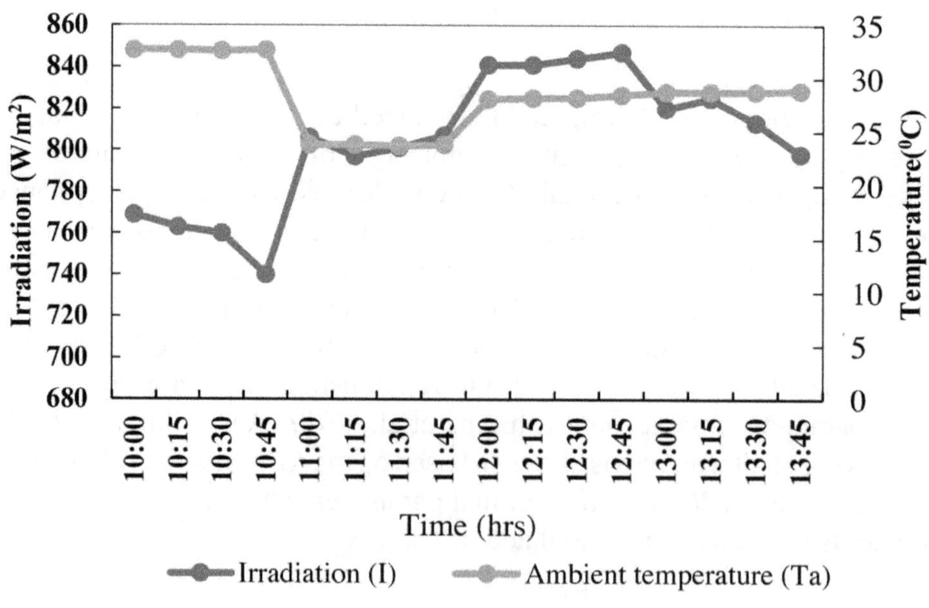

parameter because it combines the effects of operating condition and the ambient conditions in a single expression to analyse the thermal performance of the collector. As the performances are influenced by solar radiation and ambient temperature; therefore variations of these operating parameters are also monitored with respect to time. The time period considered is a part of the heating period of a characteristic solar day.

Figure 6 shows the variations of these operating/ambient parameters over time, and Figure 7 shows the variations of η and useful heat gain by the collector over the same heating period. These variations show dynamic performance (i.e. variation with respect to time) of the collector, which is also required for correct estimation of solar collector performance as the ambient parameters like solar radiation and ambient temperature are not constant over a time period of a given day. From Figure 6, it can be inferred that as the day progresses, the solar irradiation first increases and thereafter decreases till the end of the day. The variation of ambient temperature is within 23°- 33°C over the day. These variations are characteristics of the place (Bouya, et al., 2015) (Silchar, India) where the experimental readings have been taken.

Figure 7. Plot of irradiation, useful heat gain and thermal efficiency with time

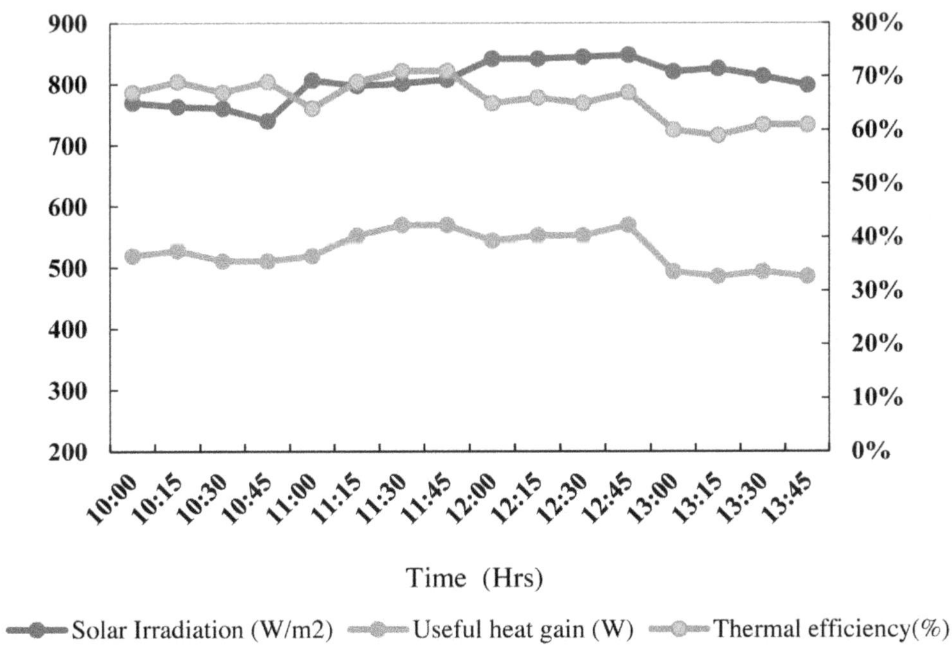

From the Figure 7, it can be observed that with the increase of solar radiation useful heat gain also increases with respect to time and then both solar radiation and useful heat gain decrease with the passage of time in the afternoon. The reason for decrease of useful heat gain is the result of the reduction in intensity of solar radiation. The thermal efficiency variation follows the same trend of the useful heat gain variation, as the latter is the output which largely affects the efficiency. The FPC attains a maximum useful heat gain of 569.29 W with a maximum thermal efficiency of 71%. The plot of η and the reduced temperature difference ($(T_i - T_a)/I$) are displayed in the Figure 8. The η values decrease with the increase of the reduced temperature difference with the maximum value of 71% obtained at a reduced temperature of 0.02. Thus, the reduced temperature containing the parameters namely, ambient temperature, water inlet temperature of the collector, and the intensity of solar radiation has direct influence on the performance of the flat plate collector. The reason for the decrease of η is the result of heat loss from the top of the collector due to the increase of average temperature of heat

Figure 8. Standard efficiency plot of a flat plate collector

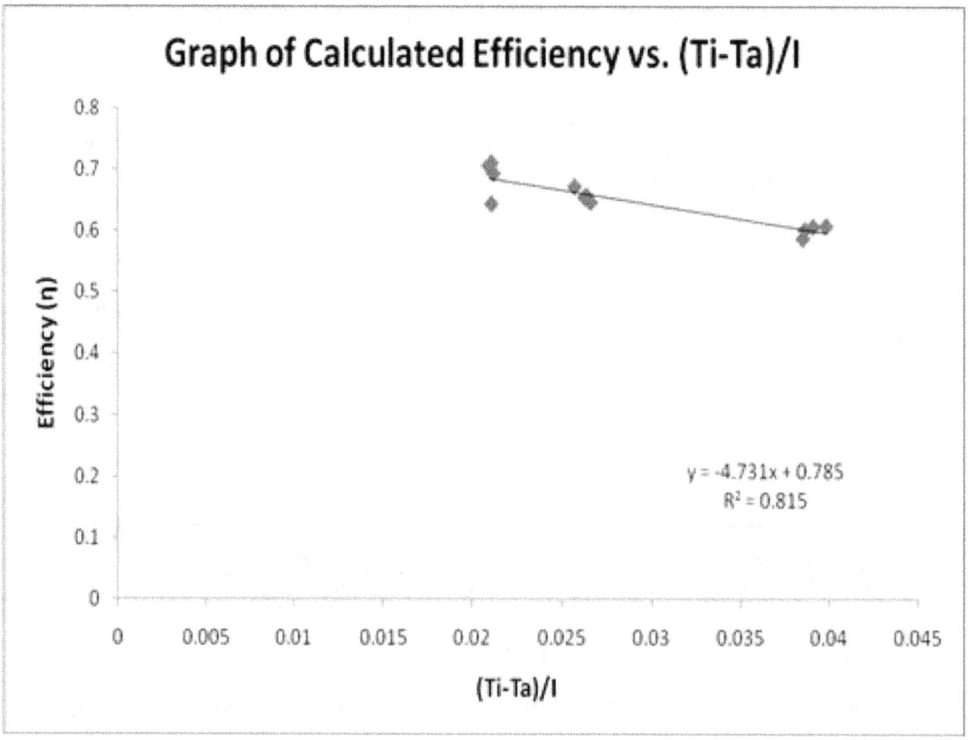

addition as the fluid temperature of the collector increases. This is one of the important characteristic curves for flat plate collector-based water heating system (Cavallaro, 2010; Cavallaro, 2009). The best fit line of Figure 8 has a degree of correlation (R-squared value) of 81.5%, which is fairly accurate. As can be seen from the plot, the product $F_R U_L$ is 4.73, which is a negative value due to negative slope and y-axis intercept, i.e. the product $F_R (\tau\alpha)e_i$ s obtained as 0.78.

Thus, from the graph of Figure 2.8, it can be concluded that $(F_R U_L) \leq$ 5.5 W/m^2 ^{0}C and $F_R (\tau\alpha)e \geq$ 0.65, which are acceptable and qualifies the tests according to the IS 12933. Figure 9 shows the variation of exergetic efficiency of the flat plate collector with respect to time duration. As can be seen from the mathematical formulation of exergetic efficiency, the same is affected by the ambient parameters as well as the outlet temperature from the collector. As the temperature difference of heat addition increases due to increased solar radiation during the noon time, there is an increase of exergy destruction, which reduces the exergetic efficiency. By comparing Figure 9 with Figure 7, it can be seen that as the solar radiation soars during 10:45 Hrs to 12:45 Hrs, the exergetic efficiency reduces to about 5.5%. Before 10:45

Figure 9. Exergetic efficiency plot of a flat plate collector

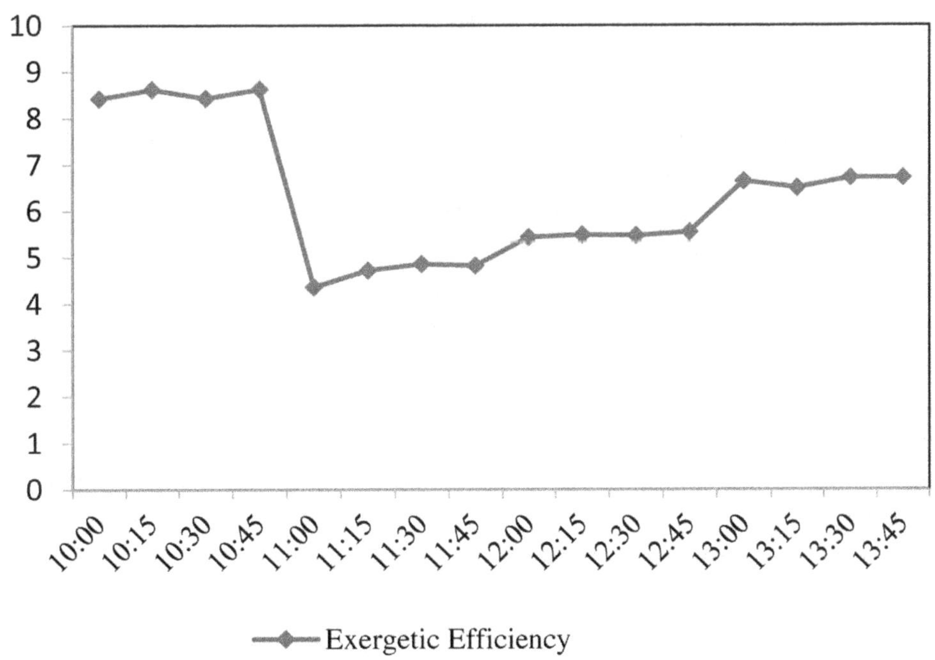

Hrs, the exergetic efficiency is higher (above 8%). Again in the afternoon period, the exergetic efficiency increases reaching about 6.7% as the intensity of solar radiation decreases. However, the exergetic efficiency cannot reach the maximum level in the afternoon (i.e. 8.61% as in 10:45 Hrs) as the useful heat gain reduces in the afternoon.

Figure 10. Standard exergetic efficiency plot of a flat plate collector

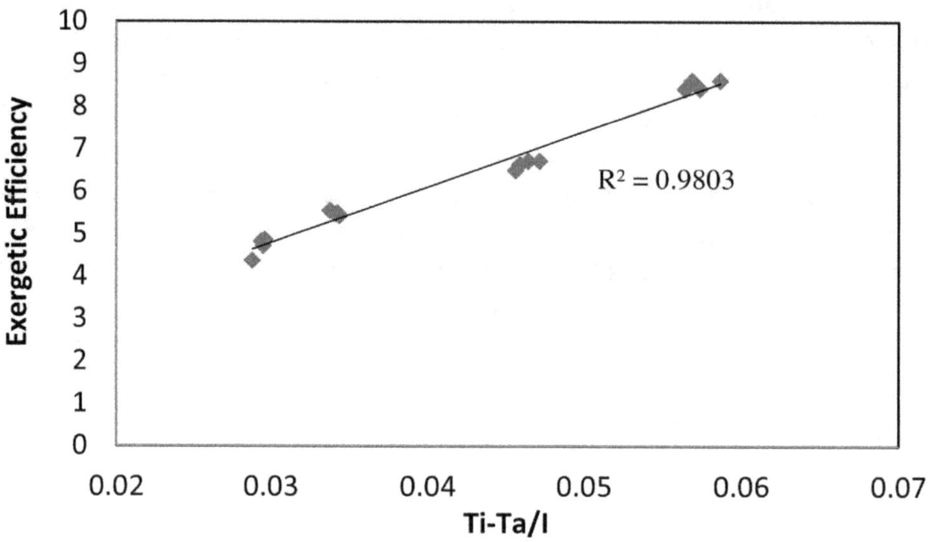

Now, Figure10 shows the standard exergetic efficiency plot of flat plate solar collector with respect to reduced temperature difference. The trend of the plot is just reverse of the standard thermal efficiency plot of the collector i.e. with increase of reduced temperature difference the second law efficiency or the exergetic efficiency increases but the first law thermal efficiency decreases. This is due to the result of heat loss from top of the collector by the increase of average temperature of heat addition resulted from the increase of reduced temperature difference, thus leading to the drop in first law thermal efficiency. However, as the reduced temperature difference increases by the decrease of solar radiation and the increase of water inlet temperature to the collector, the exergy of water increases thereby leading to the increase of exergetic efficiency of the collector. Also notwithstanding the water inlet temperature or ambient temperature, the exergetic efficiency increases due to the decrease of solar radiation leading to the decrease of the exergy destruction in addition to the decrease of the exergy of the Sun. A best fit line is shown in Figure 10, which is obtained with a degree of correlation (R-squared value) of 0.98. The maximum exergetic efficiency of 8.62% is obtained at a reduced temperature (i.e. T_i- T_a/I) of 0.058. Thus comparing the results of thermal efficiency and the exergetic efficiency with respect to reduced temperature difference, it is worth mentioning that a thermally efficient solar flat plate collector may not be an efficient solar flat plate collector in terms of exergetic performance. As the solar radiation controls the average temperature of heat addition and then the water outlet temperature from the collector and the same along with ambient temperature are weather dependent data, hence the thermal and the exergetic efficiencies can be maintained at conservative levels by controlling the water inlet temperature to the collector. This will be applicable provided the collector size is not varying.

From the previous discussion on the performance of a solar flat plate collector, it has been seen that thermal efficiency of the collector is dependent on various input operating and ambient parameters, for e.g. water inlet temperature to the collector, intensity of solar radiation, ambient temperature, etc. However, the performance of the collector is also affected by the design parameters, for e.g. collector tilt angle and the height gap between the top surface of the tilted collector and the bottom surface of the storage tank. The tilt angle can affect the natural convection of the working fluid in the riser tubes of the collector. This is because as the gravitational effect increases with the increase of tilt angle, the non-dimensional Froude number decreases. When Froude number is small, the free or natural convection is dominant. However, the reverse phenomenon happens when the tilt angle decreases. On the other hand, the height gap can affect the thermal stratification inside the storage tank, leading to a change in performance. Thus, both tilt angle and height gap need proper treatment in analyzing the thermal performance

Table 6. Effect of collector tilt angle with other operating and ambient parameters

Trail No.	Input parameters		Output parameters	
	Tilt Angle (°)	Solar Radiation (W/m^2)	Tank water Temperature (°C)	Thermal Efficiency (%)
1	32	608	43	60.4
2	38	518	53	63.3
3	41	404	48	67.8
4	44	607	40	66.3

of the collector. Therefore it is worth investigating the effects of these two design parameters on the thermal efficiency of the collector. Table 6 and Table 7 show the results of thermal efficiency with regard to different tilt angles and height gaps, respectively for various solar radiation intensities and storage tank water temperature.

Table 7. Effect of height gap with other operating and ambient parameters

Trail No.	Input parameters		Output parameters	
	Height gap (cm)	Solar Radiation (W/m^2)	Tank water Temperature (°C)	Thermal Efficiency (%)
1	17	530	33	70.18
2	26	524	35	69.85
3	35	607	36	68.23
4	44	589	48	65.76

Figure 11 and Figure 12 shows the variation of thermal efficiency with respect to the variations of collector tilt angle and height gap respectively. From Figure 11, it can be seen that with the increase of tilt angle the efficiency increases to a maximum value, which is followed by a decrease of the latter with further increase of the former. In this way, Figure 12 shows that with the increase of height gap the efficiency progressively decreases. Therefore, there is an optimum value of both tilt angle and height gap at which efficiency is the maximum. In the former case, a maximum thermal efficiency of 67.8% is obtained at an optimum tilt angle of 41°. In the latter case, a maximum thermal efficiency of 70.2% is obtained at an optimum height gap of 17 cm between the collector top surface and the storage tank bottom surface. With regards to the tilt angle, up to the optimum tilt angle variation (Figure 11), the natural convection can raise the temperature of the storage tank to higher level, which can also be corroborated from the table of tilt angle results. Although there seems to be a lack of equilibrium in the storage tank temperature due to poor thermal stratification in the storage tank, improved natural convection can circumvent the effect of improper thermal stratification up to the optimum tilt angle point.

From Figure 12, it is apparent that there is an improved thermal stratification of the stored water in the tank leading to a higher thermal efficiency than the result of tilt angle variation. However, in this case the reason for the decrease from higher thermal efficiency level (i.e. 70.2%) to lower values

Figure 11. Variation of thermal efficiency and storage tank temperature with tilt angle of the collector

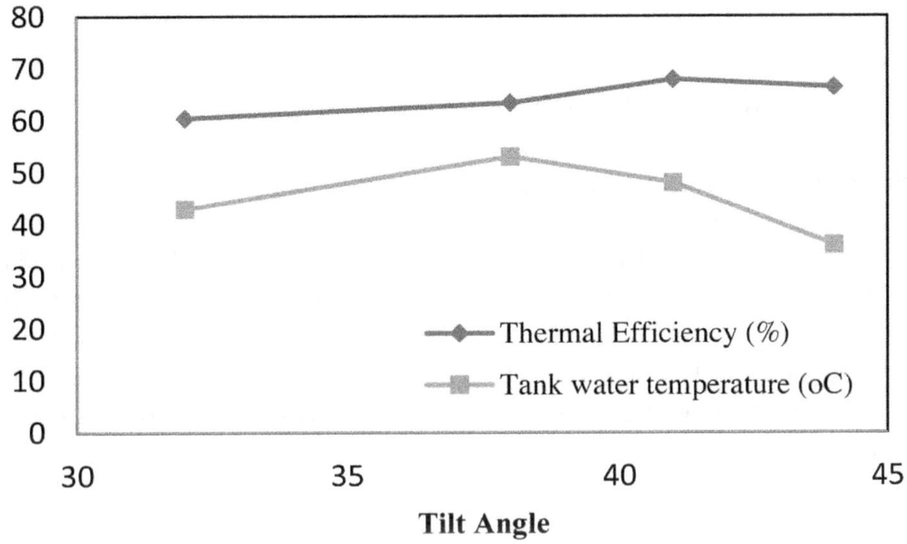

Figure 12. Variation of thermal efficiency and storage tank temperature with height gap between collector top and storage tank bottom

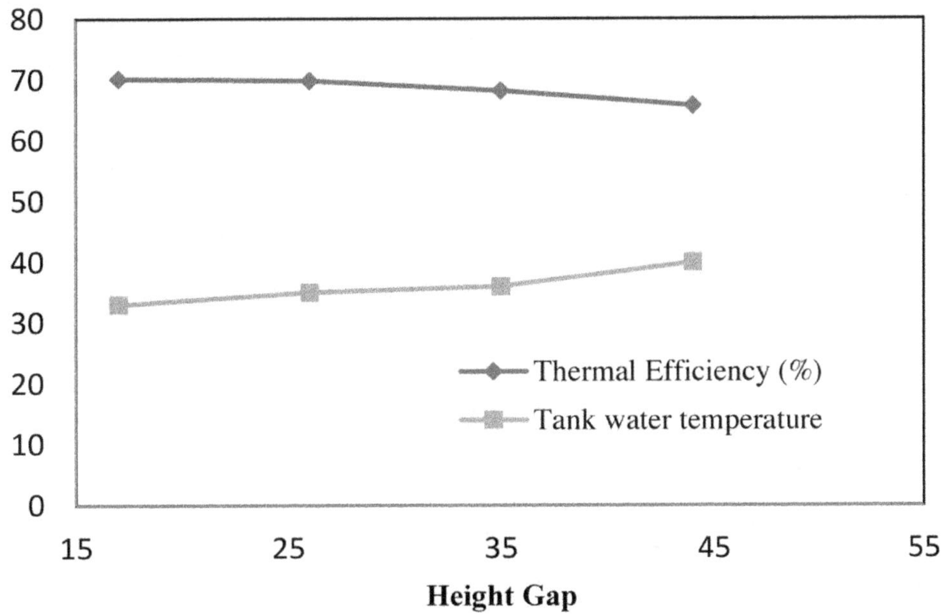

can be attributable to the lack of free or natural convection which is prevalent in the case of tilt angle variation. Further, as the storage tank temperature is higher, the thermal efficiency is also lower due to increased heat losses from the storage tank.

Optimization of SFPC Parameters

Optimization of SFPC parameters is presented using an integrated hybrid MCDM method. The optimization of SFPC is carried out based on ranking of *RPI* values and optimization results are tabulated in Table 8. The results show that trail no. 14 shows highest RPI among all the trails. The corresponding optimal setting according to the trail no. 14 is solar radiation (I) = 825 W/m^2, wind velocity (V) = 1.4 m/s, ambient temperature (T_a) = 28.8 °C and Inlet Temperature (T_i) = 66.4°C. The optimal setting provides optimal performance parameters like higher outlet temperature (T_o), higher efficiency (η), and higher

Table 8. Relative performance index values & Ranking for SFPC system

Trail. no.	Input Parameters				RPI_i values	Rank
	Solar radiation I in (W/m^2)	Wind velocity V in (m/s)	Ambient Temperature T_a in (°C)	Inlet Temperature T_i in (°C)		
1	769	1.4	32.7	76.1	0.2328	14
2	763	1.6	32.7	76.1	0.2154	16
3	760	1.2	32.6	76.2	0.2356	13
4	740	1.3	32.7	76.1	0.2151	15
5	806	2.1	23.8	47.0	0.3332	5
6	797	2.0	23.8	47.3	0.2691	10
7	801	1.9	23.7	47.4	0.2546	12
8	807	1.7	23.8	47.5	0.2599	11
9	841	1.8	28.1	57.0	0.3067	6
10	841	1.9	28.2	57.0	0.2918	8
11	844	1.6	28.2	57.0	0.2955	7
12	847	1.7	28.5	57.1	0.2739	9
13	820	1.5	28.8	66.4	0.3937	2
14	**825**	**1.4**	**28.8**	**66.4**	**0.4537**	**1**
15	813	1.1	28.8	66.5	0.3750	3
16	798	1.3	28.9	66.5	0.3711	4

useful heat gain (h), which directly or indirectly affect the performance of SFPC system (Zadeh, 1965; Brauers, et al., 2006).

Further, optimization of tank water temperature and thermal efficiency of SFPC system by varying tilt angle, solar radiation, and height gap is carried out. First, design of decision matrix is done using Eq. (2), which contains number of criteria as output parameters while number of alternatives as trail no. or input parameter settings. For this, present work considered two input parameters like title angle, and solar radiation and two response parameters such as tank water temperature (°C) and thermal efficiency (%) as output parameters. Simultaneously, 4 nos. of experiments are performed on SFPC and corresponding output parameters are determined. The experimental results as tabulated in Table 2.6 and Table 7 are considered as decision matrix in the present study. Second, normalization of the experimental data is carried out using the Eqs. (3-4). This converts the original sequence data of outlet tank water temperature and thermal efficiency into comparable sequence data in the range of 0 to 1 and the normalized results are shown in Tables 9 & 10.

Table 9. Normalized values of SFPC parameters

Trail. no.	Normalized values (N_{ij})	
	Tank water Temperature (°C)	Thermal Efficiency (%)
1	0.2337	0.2343
2	0.2880	0.2455
3	0.2609	0.2630
4	0.2174	0.2572

In the third step, determination of priority weights for each response parameters such as tank water temperature (°C) and thermal efficiency (%) is done using Fuzzy set theory. Fuzzy decision matrix is developed by taking the opinion of the DM. Opinion of four DM is taken in terms of linguistic variable using TFN as tabulated in Table 3 (Chamoli, 2015a). After that, aggregation of TFN for each of the output parameter is done to get the priority weights (ξ_i) and results are tabulated in Table 10. It is observed that the parameter tank water temperature (°*C*) yields highest priority weight ($\boldsymbol{\xi}_i$ = 0.711) results in more significance than the efficiency (η) on the SFPC performance. Simultaneously, consistency of the fuzzy decision matrix and accuracy of the priority weights (ξ_i) for each of the SFPC parameters obtained using fuzzy set theory are verified by fuzzy consistency index (CI_f) and fuzzy consistency ratio (CR_f) using Eqs. (5-6). It is seen that, the value of CI_f and $CR_{f\,as}$ obtained i.e. 0.0013 and 0.0023 < 10% results better consistency and accuracy of priority weights (ξ_i).

Table 10. Fuzzy decision matrix & priority weights of SFPC output parameters

Output response	**Triangular fuzzy weights**			**Fuzzy weights(ξ_i)**
Tank water temperature (°C)	0.227	0.428	0.471	0.711
Thermal efficiency (%)	0.163	0.272	0.433	0.289
$CI_f = 0.0013 < 10\%$; $CR_{f=}$ $0.0023 < 10\%$				

Next, development of weighted normalized decision matrix is done using Eqs. (7-8), in which priority weights (ξ_i) of tank water temperature (°*C*) and efficiency (η) are, multiplied in to each performance values of normalized decision matrix. After the weighted normalized decision matrix, determination of relative performance index values for each of the alternatives is done using the Eq. (9). This converts the multi-attribute problem into single attribute problem with considering the effect of both beneficial and non-beneficial par ammeter and interrelationship between the each criteria's/output parameter into account (Sharma et al., 2017; Acır, et al., 2017). At last, ranking of alternatives is carried out based on the *RPI* values, an alternative with highest

Table 11. Ranking of tilt angle and solar radiation for SFPC system

Trail no.	**Input parameters**		***RPI$_i$* values**	*Rank*
	Tilt Angle (°)	**Solar Radiation (W/m²)**		
1	32	608	0.4680	4
2	38	518	0.5336	1
3	41	404	0.5239	2
4	44	607	0.4746	3

RPI yields the optimal parameter setting for SFPC system and results of the same are tabulated in Table 11 and 12, respectively.

Table 12. Ranking of height gap and solar radiation for SFPC system

Trail no.	Input parameters		RPI_i values	*Rank*
	Height gap (cm)	Solar Radiation (W/m²)		
1	17	530	0.4732	4
2	26	524	0.4852	3
3	35	607	0.4858	2
4	44	589	0.5558	1

The results show that trail No. 2 in Table 11 and trail no. 4 in Table 12 show highest RPI among all the trails. The corresponding optimal setting according to the trail no. 2 is tilt angle as 38° and solar radiation as 518 W/m^2 while optimal setting for the trail no. 4 is height gap as 44 cm and solar radiation as 589 W/m^2. In both cases, optimal setting provides optimal performance parameters like higher tank water temperature, and higher efficiency (η) which directly or indirectly affect the performance of SFPC system (Zadeh, 1965; Brauers, et al., 2006).

Analysis of Variance Analysis

Further, Analysis of Variance (ANOVA) analysis with 95% of confidence interval is used to determine the significance and adequacy of the developed model. In the ANOVA analysis, the Fisher's statistical test (F-Ratio) and significance probability (P-value) followed by coefficient of determination (R^2) and adjusted-R^2 are determined to test the sufficiency and fitness of the proposed model. The F-value denotes the most influencing parameters on the performance characteristics while P-value to show the statistical significance of process variables of SFPC system. During the analysis, process variables which have larger F-Values and lower P-values ($P<0.05$) signify the statistically significant parameters (Kulekci et al. 2011). The R^2 signifies the total variation in the response expected by the model. The purpose of adjusted-R^2 is to analyze the model adequacy and fitness. In this research stepwise elimination method is used to remove the insignificant parameters to adjust the fitted quadratic model (Rao, & Kalyankar, 2014). In ANOVA

Table 13. ANOVA for RPI values of SFPC system

Source	DF	Adj SS	Adj MS	F-Value	P-Value
Regression	2	0.03178	0.015891	5.06	0.024
Solar radiation (I)	1	0.02978	0.029784	9.48	0.009
Wind velocity (V)	1	0.0115	0.011505	3.66	0.008
Error	13	0.04084	0.003141		
Total	15	0.07262			
R^2=96.97%, R^2(adj)=92.14%, R^2 (Pred)= 95.77%					

analysis, results as tabulated in Table 6 are taken and corresponding ANOVA results are shown in Table 13.

The ANOVA results show that the parameter solar radiation (I) and wind velocity (V) are the most significant factors influencing the assessment of RPI values of SFPC performance. The parameters Ambient Temperature (T_a) and Inlet Temperature (T_i) are considered as statistically insignificant, since they failed to test the significance. Subsequently, the P values for solar radiation (I) and wind velocity (V) are found to be less than 0.05, which indicate that the model and associated terms are statistically significant (Konig, et al., 1985). The other parameters except parameter highlighted in Table 13 are considered to be statistically insignificant. The values of R^2 is calculated to be 0.9697 for RPI values, it implies that 95% of experimental data are well-fitted. Higher the R^2 coefficient gives the satisfactory model to the experimental data. Finally, the adequacy and fitness of the model is calculated by adjusted-R^2 values. The high value of adjusted-R^2 (0.9214 for RPI values) supports a high correlation between the experimental and the predicted values (Kumar, et al. 2014).

Confirmatory Test

Additionally, confirmatory tests are conducted to verify or validate the optimal setting obtained via the proposed method. The optimal setting i.e. solar radiation (I) = 825 W/m^2, wind velocity (V) = 1.4 m/s, ambient temperature (T_a) = 28.8 °C and Inlet Temperature (T_i) = 66.4°C is used for confirmatory experiments and the corresponding results are shown in Table 14. The results show that confirmatory tests results are comparable and acceptable with experimental results for the optimal conditions.

Table 14. Confirmatory tests for SFPC thermal system

Input Conditions	Output Parameters	Experimental results	Confirmatory results
Test No-14: $I = 825$ W/m², $V = 1.4$ m/s, $T_a = 28.8$ °C and $T_i = 66.4$°C	T_o in °C	60.6	62.0
	η in %	58.86	59.92
	h in W	485.576	486.521
	η_{Ex} in %	6.495	6.532

SUMMARY

Modeling and optimization of SFPC is presented considering four input parameters like Solar radiation (*I*), wind velocity (*V*), ambient temperature (T_a) and inlet temperature (T_i) and two response parameter such as outlet temperature (T_o), efficiency(η), and useful heat gain (h) as output parameters. An integrated hybrid MCDM method i.e. fuzzy set theory with MOOSRA is illustrated. Optimization results reveal that trail No 14 gives highest *RPI* among the other trail no for the SFPC. The optimal operating/input parameters obtained are solar radiation (*I*) = 825 W/m², wind velocity (*V*) = 1.4 m/s, ambient temperature (T_a) = 28.8°*C* and Inlet Temperature (T_i) = 66.4°C, which give most optimal response/output parameters such as higher outlet temperature (T_o), higher Efficiency (η) and higher heat gain (h) that directly or indirectly affect the performance of SFPC system.

In addition, optimizations of other parameters like tank water temperature and efficiency considering the four independent parameters have also been done. The results show that trail No. 2 for Table 11 and trail no. 4 for the Table 12 show highest RPI among all the trails. The corresponding optimal setting according to the trail no. 2 is tilt angle as 38° and solar radiation as 518 W/m² while optimal setting for the trail no.4 are height gap as 44 cm and solar radiation as 589 W/m². Furthermore, parametric analysis shows that operating parameters like solar radiation, ambient temperature, and water inlet temperature of the collector are most significant on the SFPC performance and optimal values of outlet temperature (T_o), efficiency (η) and useful heat gain (h) are obtained as 60.6 °C, 58.86%, and 485.576 W, respectively.

Additionally, ANOVA analysis is carried out to determine the significance and adequacy of the developed model. The results show that solar radiation and wind velocity are the major parameters and the model fitted data are well fitted with experimental result within 95% confidence level. At last, confirmatory tests show comparable and acceptable results with experimental values. Finally, as the MCDM technique used give optimized values of the considered design and operating parameters and efficiency of the collector, and as these are obtained with good accuracy and with statistical significance, the obtained optimized values can be practically implemented in the present solar collector. Therefore, it is concluded that hybrid technique is feasible in modeling and optimizing the SFPC parameters and can be used as systematic framework model for modeling and optimization of other thermal systems also.

REFERENCES

Acır, A., Emin, M., & Cakırog, R. (2017). Parametric optimization of energy and exergy analyses of a novel solar air heater with grey relational analysis. *Applied Thermal Engineering*, *122*, 330–338. doi:10.1016/j.applthermaleng.2017.05.018

Athanasios, K., Varvara, M., Estivaliz, L. M., & Konstantinos, S. (2016). A Comparative Study of Multiple-Criteria Decision-Making Methods under Stochastic Inputs. *Energies*, *9*(566), 1–21.

Bhowmik, C. (2014). Optimization of Process Parameter using Theory of Constraints. *International Journal of Basic Applied Science and Research*, *1*, 7–10.

Bouya, M., El Majd, B. A., Agdid, M., & Abdellah, A. B. (2015). Solar Heater Technologies Selection Using Multi-criteria Analysis. *Journal of Energy and Power Engineering*, *9*(4), 393–398.

Brauers, W. K. M., & Zavadskas, E. K. (2006). The MOORA method and its application to privatization in a transition economy. *Control and Cybernetics*, *35*(2), 445–469.

Cavallaro, F. (2009). Multi-criteria decision aid to assess concentrated solar thermal technologies. *Renewable Energy*, *34*(7), 1678–1685. doi:10.1016/j.renene.2008.12.034

Cavallaro, F. (2010). Fuzzy TOPSIS approach for assessing thermal-energy storage in concentrated solar power (CSP) systems. *Applied Energy*, *87*(2), 496–503. doi:10.1016/j.apenergy.2009.07.009

Chamoli, S. (2015). Hybrid FAHP (fuzzy analytical hierarchy process)-FTOPSIS (fuzzy technique for order preference by similarity of an ideal solution) approach for performance evaluation of the V down perforated baffle roughened rectangular channel. *Energy*, *84*, 432–442. doi:10.1016/j.energy.2015.03.007

Chamoli, S. (2015). Preference selection index approach for optimization of V down perforated baffled roughened rectangular channel. *Energy*, *93*, 1418–1425. doi:10.1016/j.energy.2015.09.125

Dorfeshan, Y., Mousavi, S. M., Mohagheghi, V., & Vahdani, B. (2018). Computers & Industrial Engineering Selecting project-critical path by a new interval type-2 fuzzy decision methodology based on MULTIMOORA, MOOSRA and TPOP methods. *Computers & Industrial Engineering*, *120*, 160–178. doi:10.1016/j.cie.2018.04.015

Konig, W., Wulf, C., & Willerscheid, H. (1985). Machining of fibre reinforced plastics, cirp annals. *Manufacturing Technology*, *34*, 537–548.

Kulekci, M.K., Esme, O. E., & Kazancoglu, Y. (2011) 'Modeling and prediction of weld shear strength in friction stir spot welding using design of experiments and neural network. *Materials Science and Engineering Technology*, *42*, 990-995.

Kumar, R., Kumar, K., & Bhowmik, S. (2014). Optimization of mechanical properties of epoxy based wood dust reinforced green composite using taguchi method. *International Conference on Advances in Manufacturing and Materials Engineering*, *5*, 688-696. 10.1016/j.mspro.2014.07.316

Kuo, C. J., Su, T., Jhang, P., Huang, C., & Chiu, C. (2011). Using the Taguchi method and grey relational analysis to optimize the fl at-plate collector process with multiple quality characteristics in solar energy collector manufacturing. *Energy*, *36*(5), 3554–3562. doi:10.1016/j.energy.2011.03.065

Lahu, P. M., Gandigude, A. U., & Kale, S. A. (2014). Selection of Material for Wind Turbine Blade by Analytic Hierarchy Process (AHP) Method. *Applied Mechanics and Materials*, *612*, 145–150. doi:10.4028/www.scientific.net/AMM.612.145

Rao, R. V., & Kalyankar, V. D. (2014). Optimization of modern machining processes using advanced optimization techniques—A review. *International Journal of Advanced Manufacturing Technology*, *73*(5-8), 1159–1188. doi:10.100700170-014-5894-4

Sharma, A., Chauhan, R., Singh, T., Kumar, A., Kumar, R., & Sethi, M. (2017). Optimizing discrete V obstacle parameters using a novel Entropy-VIKOR approach in a solar air flow channel. *Renewable Energy*, *106*, 310–320. doi:10.1016/j.renene.2017.01.010

Suresh, K. B., Subbaraju, N. V., Reddy, S. M., & Nageswara, R. D. (2006). The material selection for typical wind turbine blades using a MADM approach & analysis of blades. *Conference on MCDM*, 1-11.

Ummadisingu, A., & Soni, M. (2011). Concentrating solar power–technology, potential and policy in India. *Renewable & Sustainable Energy Reviews*, *15*(9), 5169–5175. doi:10.1016/j.rser.2011.07.040

Zadeh, L. A. (1965). Fuzzy sets. *Information and Control*, *8*(3), 338–353. doi:10.1016/S0019-9958(65)90241-X

Chapter 3
Modeling and Optimization of Concentrated Solar Thermal System

ABSTRACT

This chapter discussed the modeling and optimization of concentrated solar thermal (CST) system using data envelopment analysis-based ranking (DEAR) method. Experimentation on CST under different environment (i.e., summer and winter season) of Silchar (Southern Assam, India) conditions is done. In order to improve the temperature (T), height of the receiver, that is, summer height and winter height (hs and hw) for summer and winter conditions, CST parameters are optimized. Study concludes that performance of CST under considered environments is feasible and use of Data envelopment analysis based ranking (DEAR) method found feasible and provides optimal results. Additionally, ANOVA analysis is carried out to determine the significance and adequacy of the developed model. The results show that all the response parameters are the major parameters for CST system and also model fitted data well fitted with experimental result within the 95% confidence level. At last, confirmatory test are conducted to verify and validate the proposed method with the experimental results.

DOI: 10.4018/978-1-7998-3523-3.ch003

INTRODUCTION

Concentrated solar thermal (CST) system is a modern development of solar thermal collector for thermal power generation and also process steam generation for cooking. CST systems are of two types- line focussing and point focussing systems (Ummadisingu & Soni, 2011). There are mainly two types of line focussing CST systems (Desai & Bandyopadhyay, 2017); one is the linear fresnel reflector (LFR) and the other is the parabolic trough collector (PTC). Both the systems concentrate the direct solar radiation onto a linear receiver (Purohit, et al. 2013). The thermal modeling and optimization of PTC shall be undertaken in the subsequent chapter of this book. The point focussing CST systems have larger concentration ratio than the line focussing CST system, and it is able to raise the fluid temperature inside the receiver to a very high level. However, cost of the point focussing system is increased due to the increase of concentration ratio. The point focussing system is categorised into two major types. One is the central receiver system (CRS) and the other is the parabolic dish reflector (PDR). Both the systems play a significant role in utilizing the solar radiations for power production (Romero & Augilar, 2013). The CRS, a popular point focussing system, has large potential. It plays a considerable role in utilizing the solar radiations for thermal applications in power generation and also steam cooking. Once the solar radiant flux is focussed onto the receiver by the CRS, steam can be generated at a very high temperature and pressure, which can then be passed through a steam turbine to produce solar thermal power.

On the other hand, in case of PDR, a reversible Stirling engine can directly be placed at the receiver location, which can convert the thermal energy directly into solar thermal power. Photographs of central receiver system (of different orientations) are shown in Figure 1 and Figure 2. The major components are a field of heliostats and a receiver unit (i.e. the solar thermal collector), which is mounted at the focal point of the heliostat mirrors (Figure 2a). Heliostats are devices that include mirror (Figure 2b), usually plane mirror, which turns so as to keep reflecting sunlight toward a predetermined target, compensating for the Sun's apparent motions in the sky. Heliostats are usually controlled by computers. The computer is given the latitude and longitude of the heliostat's position on the earth and the time and date. From these, using astronomical theory, it calculates the direction of the Sun as seen by the mirror. Then, according to the direction of the target, the computer calculates the direction of the required angle-bisector, and sends control signals to motors, often stepper

Figure 1. Photograph of the central receiver on top of the tower along with heliostats on field (Courtesy- National Institute of Solar Energy, Gurugram, India)

motors, so they turn the mirrors to the correct alignment. This sequence of operations is repeated frequently to keep the mirror properly oriented. The fabrication of the mirrors, their stands involve lot of craftsmanship, so that it optics support reflection of the sunlight onto the receiver unit.

Figure 2a. Arrangement of heliostats around a central receiver

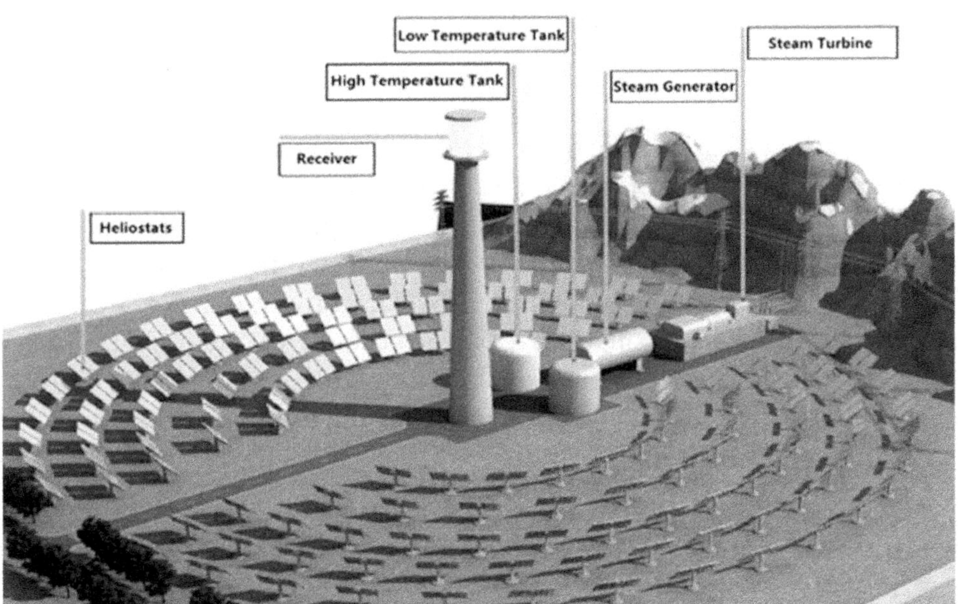

Figure 2b. Heliostat plain mirror and its stand

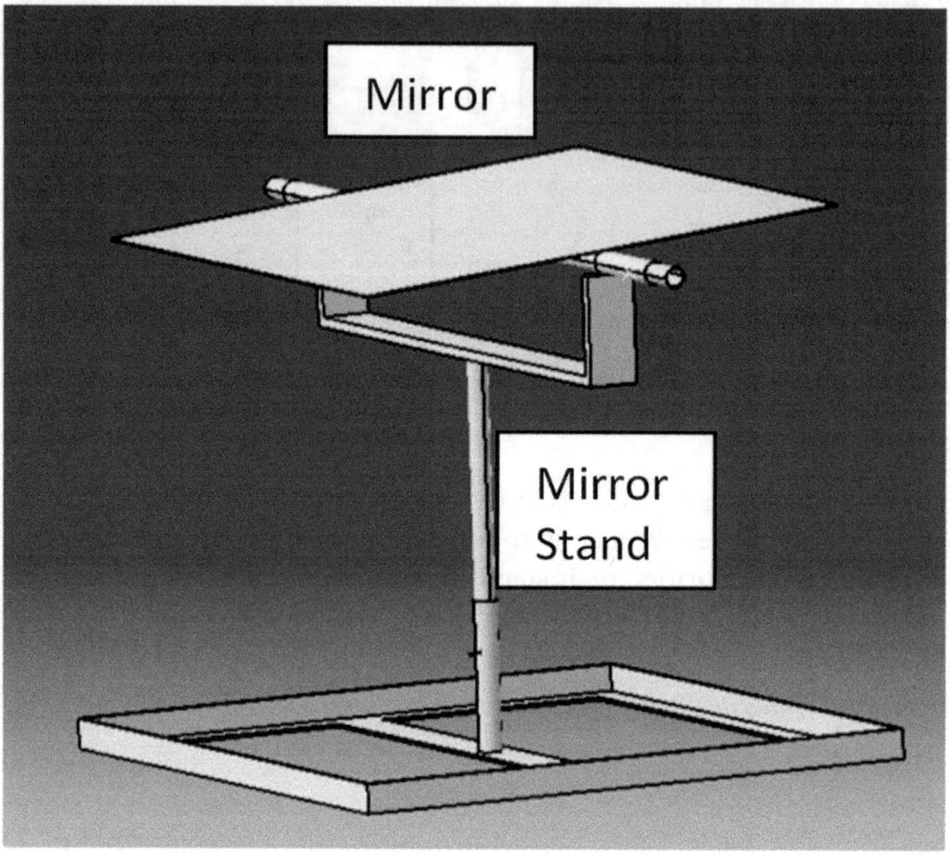

Figure 3 shows a schematic of the ray diagram concentrating radiation on the central receiver. Here L is the distance of the receiver from a heliostat mirror, h is the height of the receiver, c is the height of the mirror up to its mid plane, θ is the angle of incident or the reflected ray, β is the heliostat tilt angle, and α is the altitude angle. The height of the receiver can be theoretically calculated using Equation (1). The incidence angle θ can be calculated using analytical formula as given in Equation (2). And a derived angle i (refer Figure 3) can be calculated using the formula in Equation (3).

$$h = -L\tan(\alpha + 2\beta) \tag{1}$$

Figure 3. Ray Diagram

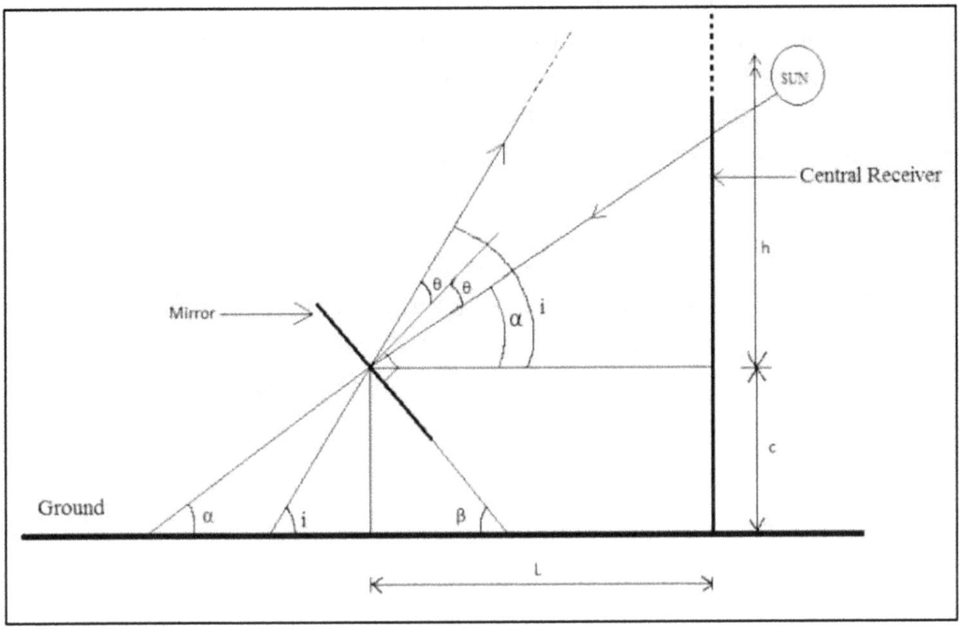

$$\theta = \left| \frac{\alpha - \tan^{-1}\left(\frac{h}{L}\right)}{2} \right| \tag{2}$$

$$i = 2\theta + \alpha \tag{3}$$

Apart from the design of CST, the performance analysis of CST is also important to know behaviour of system. In actual, any system including CST, the performance is depending on its operating/input parameters and performance/output parameters. Various input parameters like altitude angle (α), length (L), heliostat tilt angle (β) affects the performance of CST. The height of the receiver and outlet temperature (T) of the receiver are directly related to the efficiency of the CST. Improper selection of any of these parameters leads to reduction in temperature (T), as well as reduce the performance and efficiency of the CST system. A detailed energy and exergy analysis are however needed to understand the efficacy of the outlet temperature on the thermal and exergetic efficiencies of the CST system. In the solar receiver, the heat flux arriving at the aperture area of the heliostat mirror is transferred into a heat flow directed to the heat transfer fluid, which is considered water in the present study. The heat is used to raise the temperature of water outlet from the receiver (T). However, on the way from the aperture plane to the fluid inside the receiver, some optical and thermal losses occur, which can be modelled by using the following expression:

$$Q_{inc} = Q_{abs} + Q_{loss} \tag{4}$$

where Q_{inc}=incident heat flux due to solar radiation on the receiver, Q_{abs}=heat absorbed by the receiver fluid, i.e. water, Q_{loss}=heat losses due to convection (Q_{conv}), radiation (Q_{rad}) as well as optical losses ($Q_{optical}$), i.e.

$$Q_{loss} = Q_{rad} + Q_{conv} + Q_{optical} \tag{5}$$

The expressions for these entire heat transfer equations can be written in the following manner-

$$Q_{abs} = mc_p(T - T_i) \tag{6}$$

where 'm' is the mass flow in the receiver, c_p is the specific heat of water (i.e. 4.18 kJ/kg-K), T is the outlet temperature of water, and T_i is the water inlet temperature to the receiver, which is taken at the ambient condition. The radiative heat transfer equation to the heat transfer fluid, i.e. water in the receiver can be expressed as-

$$Q_{rad} = \varepsilon\sigma\left(T_r^4 - T^4\right)A_r \tag{7}$$

where ε is emissivity of copper receiver tube (which lies in the range 0.08 – 0.12), σ is the Stefan's Boltzmann constant, Tr is the receiver surface temperature, and Ar is receiver surface area. The convective heat transfer to the working fluid can be expressed as-

$$Qc_{onv} = hw_{(}\mathrm{T}r - T)\mathrm{A}r \tag{8}$$

where h_w is the convective heat transfer coefficient, which can be expressed as-

$$h_w = 5.7 + 3.8V \tag{9}$$

where V is the wind speed, which can be assumed to be constant at 6 kmph for the region where the present study is conducted. The best optical efficiency of the present setup was found to be 0.8 for the existing setup at the Regional Test Center, NIT Silchar, Assam, India. Hence, the optical loss ($Q_{optical}$) was 0.2 times the incident heat flux. Now the thermal efficiency based on the above first law analysis can be expressed in the following manner-

$$\eta = \frac{Q_{abs}}{Q_{inc}} \tag{10}$$

As the first law analysis can only provide the maximum attainable thermal efficiency of the CST system, the work potential of the heat transfer process from receiver to the working fluid can only be analysed with the help of second law analysis, i.e. the exergy analysis. Similar to energy balance equation, exergy balance equation can be expressed as-

$$Ex_{inc} = Ex_{abs} + Ex_{loss} \tag{11}$$

where Ex_{inc} is the exergy associated with incident heat flux on to the receiver, which is equal to the exergy of the heat absorbing fluid (Ex_{abs}) and exergy loss (Ex_{loss}). Exergy losses include that associated with convection, radiation, and optical losses. The inlet exergy of the incident heat flux can be expressed as-

$$Ex_{inc} = Q_{inc}\left(1 - \frac{Ta}{T_{Sun}}\right) \tag{12}$$

where T_{Sun} is the Sun temperature, which is around 5762 K, and T_a is the ambient temperature. The exergy of the heat absorbing fluid can be expressed as-

$$Ex_{abs} = mc_p\left(T_O - T_i\right) - mc_pT_a ln\left(\frac{T_o}{T_i}\right) \tag{13}$$

Exergy loss can be expressed as-

$$Ex_{loss} = Ex_{inc} - Ex_{abs} \tag{14}$$

Therefore, the exergetic efficiency or the Second law efficiency can be expressed as-

$$\eta_{Ex} = \frac{Ex_{abs}}{Ex_{inc}} \tag{15}$$

Optimization of CST parameters is required to achieve better outlet temperature (T) and receiver height (h). Optimized values of these parameters shall ensure better thermal performance of the CST system in terms of first law and second law efficiencies. Further, optimization of CST system is considered to be MCDM problems. Hence, suitable MCDM method needs to be used for modeling and optimization of CST system. Many investigators have worked on the design and analysis of different solar energy systems (Mohanty, et al., 2017; Mehrabi, et al. 2017; Kadek, et al. 2016; Mevlut, 2013) other than CST. Mohantly et al. (2017) used ANFIS based model for modeling and optimization of solar water heating system. Mehrabi et al.(2017) employed fuzzy based model for prediction of irradiative enclosure. Kadek, et al. (2016) applied Fuzzy-AHP based model for modeling and optimization of site selection for solar farm and Mevlut, (2013) utilized AHP alone for site selection of solar farms-based on GIS. It is observed that the concerned authors (Mohanty, et al., 2017; Mehrabi, et al. 2017; Kadek, et al. 2016; Mevlut, 2013; Suman et al., 2019; Cavallaro, 2009; Nixon, 2013; Aragonés-Beltrán, et al. 2014) have ignored the considerations of criteria weights as well as beneficial and non-beneficial parameters into account. Therefore, all the conflicting criteria in the decision-making process might not have been entailed by the authors. Most of the studies were concentrated on site selection for solar system. None of the literature has done the modeling and optimization of CST systems using MCDM method. Hence, the present chapter has put forward to show the application of an MCDM technique called DEAR method (Muthuramalingam, et al. 2014) for modeling and optimization of CST system. DEAR method offers many advantages like ability to solve multi-objective problem, considers multiple criteria with subjective and objective data during optimization, easy to maintain the consistency, less computational time required, very simple and stable technique, and minimum mathematical calculation involved to the decision makers.

Experimental Details

As heliostats are used for concentrating solar radiation, a detailed study and a series of experiments on the heliostat mirrors are required to be performed on different days of summer and winter in order to determine the height of the heliostat mirrors for concentration of the reflected solar radiation from the mirrors. Thus, first stage of experimentations is concentrated on determining the summer height and winter height with respect to the changing altitude angle with the season. The receiver is kept at a distance of 1.5m from the axis of the heliostat mirrors, which is not changed during experimentations. Similarly, the tilt angle of the mirror is also kept fixed at 32.8°, which is equal to the latitude of the place (Silchar, India) plus 8°. Given the altitude and the solar incidence angle, height of the receiver system and the angle of

Table 1. Height of receiver of CST system [summer & winter season]

Trail No.	Altitude Angle Summer (α)	Altitude Angle Winter (α)	Length of mirror (m)	Different title angle β(deg.)	Summer height of receiver (h_s)	Winter height of receiver (h_w)
1	70.2	42.2	1.5	32.8	6.107	4.672
2	70.6	42.3	1.5	32.8	5.928	4.644
3	70.9	42.4	1.5	32.8	5.800	4.617
4	71.3	42.5	1.5	32.8	5.637	4.589
5	71.7	42.6	1.5	32.8	5.483	4.562
6	72.1	42.7	1.5	32.8	5.336	4.536
7	72.5	42.8	1.5	32.8	5.197	4.509
8	72.8	43	1.5	32.8	5.097	4.457
9	73.2	43.1	1.5	32.8	4.968	4.432
10	73.6	43.3	1.5	32.8	4.846	4.381
11	73.9	43.4	1.5	32.8	4.757	4.356
12	74.3	43.6	1.5	32.8	4.644	4.307
13	74.7	43.8	1.5	32.8	4.536	4.259
14	75	43.9	1.5	32.8	4.457	4.236
15	75.4	44.1	1.5	32.8	4.356	4.189
16	76.1	44.3	1.5	32.8	4.189	4.144
17	76.4	44.5	1.5	32.8	4.121	4.099
18	76.8	44.7	1.5	32.8	4.033	4.055
19	77.1	44.9	1.5	32.8	3.970	4.012

reflection of rays onto the receiver can be determined using the Equations (1-3). The data generated is given in Table 1.

After setting the receiver at its required height, the temperature of the working fluid in the central receiver is recorded over a given period of time. The initial temperature of the water is kept fixed at the ambient temperature of the day. The experimental setup of the central receiver system is shown in Figure 4. A total 6 heliostats are considered as shown. On different days, the inclinations of the heliostats are also manoeuvred to ensure reflection of the sun rays from the mirrors onto the receiver. The experimental datasets generated of the tilt angles of the individual mirrors and the receiver temperature are given in Table 2. These datasets are generated during three days of experimentations at the Regional Test Center at Silchar, India. Six heliostat mirrors are considered, whose stands are oriented in a suitable manner around the central receiver and the temperature readings of the working fluid, i.e. heat transfer fluid (water) after equal interval of time for a particular day are noted. The global solar radiation is measured in the same plane of the heliostat mirrors using a solar pyranometer (range 0-1500 W/m^2 with an accuracy ± 1 W/m^2). The water temperature is measured using PT 100 RTD type sensor (range -40 to 60°C with an accuracy ± 0.2°C). For measuring flow rate, an electronic flow meter is used (range 0.5-25 lpm, accuracy ± 1%).

Figure 4. Experimental setup for taking the temperature readings (receiver top is open)

Table 2. Temperature readings for different β values of different mirrors of CST system

Trail No.	Time (t)	Global solar radiation (G) in W/m²	Tilt angle (°)						Temperature (T) in °C
			Mirror-1 (β_1)	Mirror-2(β_2)	Mirror-3(β_3)	Mirror-4(β_4)	Mirror-5(β_5)	Mirror-6(β_6)	
Set-1									
1	11:15:00	828.33	40	35	35	27	27	33	27
2	11:30:00	828.821	37	35	36	25	28	30.5	31
3	11:45:00	810.657	38.4	34.5	32	25	27.3	32	34
4	12:00:00	806.379	36	33	30	22.5	25	31	36
5	12:15:00	778.724	35	35.5	32.5	26	30	34	39
6	12:30:00	768.954	38.2	35	33	26	31	36.5	40
7	12:45:00	752.867	38	37	34	29	33	36	42
8	1:00:00	712.318	41.5	38	34	31	34.5	37.5	45
9	1:15:00	666.563	41	40	36	34	37.5	42.5	46
10	1:30:00	640.563	43	41	39	36	37.5	44	45
Set-2									
1	11:15:00	205.384	35.5	34.5	35	26.5	26	31	27
2	11:30:00	205.393	34	33	33.5	25	26,3	30.5	29
3	11:45:00	206.557	33	34	32	23	27	32	31
4	12:00:00	205.534	32	36	30	22	28	33	32
5	12:15:00	205.434	35	35.5	32.5	26	30	35	35
6	12:30:00	206.465	38	35	33	28	32	36	38
7	12:45:00	207.039	38	36.5	34	29	33.5	37	42
8	1:00:00	206.789	41	39	34.5	31	35	39	44
9	1:15:00	207.794	41	39	36	34	37.5	43	41
10	1:30:00	208.102	43	41	39	36	37	44.5	40
Set-3									
1	11:50:00	862.877	36	33	33	28	30	34	27
2	12:05:00	840.519	32	34	31	27	31	30	41
3	12:20:00	825.324	39.5	34	32	28	32	35	50
4	12:35:00	810.296	41.5	35	30	20	33	41	59
5	12:50:00	785.048	41	35.5	32.5	29	35	40	66
6	1:05:00	768.954	41	35	30	36	37.5	42.5	68
7	1:20:00	752.867	44	36	32	30	37	45	72
8	1:35:00	712.318	44	38	32	33	39	47	77
9	1:50:00	666.563	46	40	33	35	42	49	81
10	2:10:00	585.034	48	40	30	35	43.5	50	76
11	2:25:00	466.907	50.5	43	32	37	41.5	53	77
12	2:40:00	223.194	52	44	32	40.5	49	54	73
13	2:55:00	365.86	48.5	42.5	31.5	42.5	50	55	69
14	3:10:00	376.325	55	52	34	44	52	57	64
15	3:25:00	328.193	55	52	35	48	57	60	61
16	3:35:00	255.265	60	51	35	54	60	62	57
17	3:50:00	253.964	62	53	37	55	62	64	53
18	4:05:00	222.367	62	55	38	57	62	65	51

DEAR Method

In this section, an MCDM technique is discussed for modeling and optimization of CST system. The schematic flow chart of the proposed method is shown in Figure 5. The details about the DEAR (Suman et al., 2019) method is explained as follows:

Step-1: Determination of weights

The DEAR method starts with first determination of parameter weights. The weight for each of the parameters is determined by the ratio of parameter at any trail no. to the sum of all parameters. The computation to determine the weights of each parameter is done using the following equations.

For higher-the-better parameter

$$\xi_h = \frac{\chi_i}{\sum_{i=1}^{n} \chi_i} \text{ For } i = 1, 2, \ldots\ldots n \tag{4}$$

For lower-the-better parameter

$$\xi_l = \frac{1/\chi_i}{\sum_{i=1}^{n} 1/\chi_i} \text{ For } i = 1,2, \ldots\ldots n \tag{5}$$

where, ξ_h is the weights of higher the better parameters for the *i*th parameter to corresponding *j*th parameter, ξ_l is the weights of lower the better parameters for the *i*th parameter to corresponding *j*th parameter χ denotes the performance parameter value of *i*th trail.

Step-2: Weighted decision matrix

In this step, weighted decision matrix is formulated by multiplying individual output parameter with weight of each parameter (Cavallaro, 2009). This step is necessary because it considers the weights effect, because different weight fractions have different effects on the results. The weighted decision matrix is formulated using Equation (5).

$$\Psi_1 = [(\xi h \times \chi ij_{)]} \text{ for } i=1,2,.....n \ \& \ j=1,2,........m \qquad (6)$$

$$\Psi 2 = [(\xi l \times \chi ij)_{]} \text{ for } i=1,2,.....n \ \& \ j=1,2,........m \qquad (7)$$

where, $\Psi 1$ is the weighted decision matrix for higher the better parameters, $\Psi 2$ denotes the weighted decision matrix for lower the better parameters, n *i*s the number of trials and m *d*enotes the number of parameters.

Step-3: Evaluation of Multi-Performance Ranking Index (MPRI)

Figure 5. Schematic flow chart of DEAR method for CST

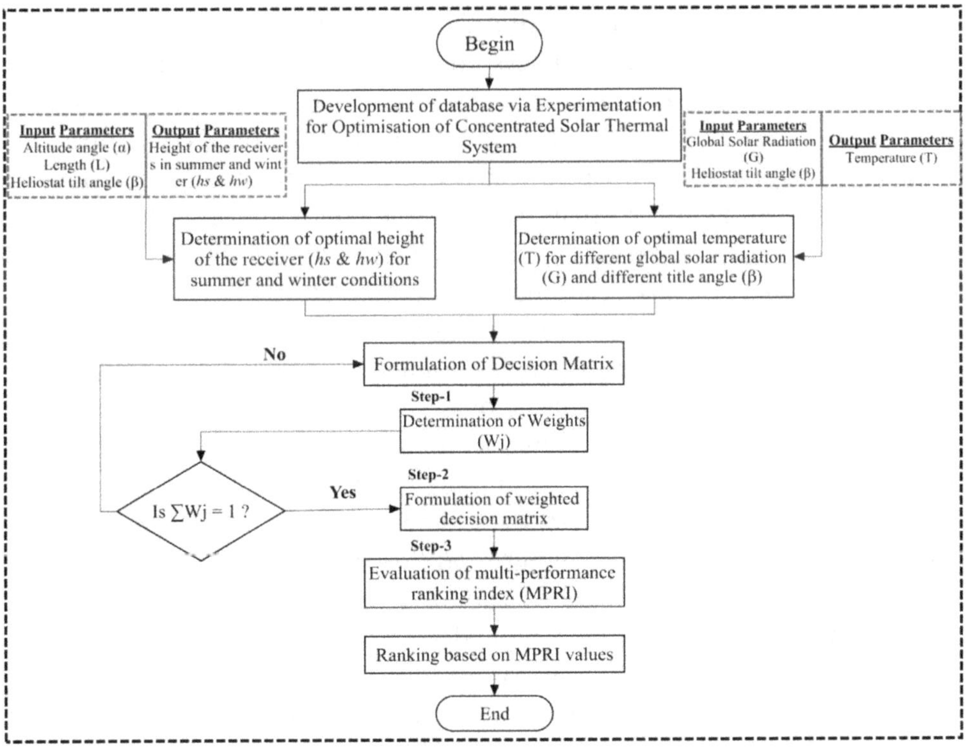

After determination of weighted decision matrix, MPRI values for each of the output parameters is evaluated using Equation (6). This converts the multi-response to the single response with consideration of weight effect in account. At last, ranking is done based on the MPRI values (Suman et al.,

2019 & Nixon, 2013) to get the optimal parameters for CST system. The highest MPRI value yields the optimal setting for CST which provides optimal output parameters for CST system.

$$MPRI_i = \frac{\psi_1}{\psi_2} \tag{8}$$

where, MPRI represent the multi-performance ranking index for each ith trail number for system.

Modeling of CST System

The modeling of CST under different environment [i.e. summer and winter] of Silchar, Assam, India is done using DEAR method (Suman et al., 2019). Work optimized the CST parameters under two conditions. First- optimal height of the receiver (h_s & h_w) for summer and winter conditions. Second- optimal temperature (T) for different global solar radiation (G) and different tilt angle (β) of six numbers of mirrors. For the first optimization, height of the receivers (h_s and h_w) is taken as output parameter while altitude angle (α), length (L), heliostat tilt angle (β) are taken as input conditions. Similarly, for the second optimization, parameter i.e. temperature (T) is considered as output parameter while global solar radiation (G) and different title angle (β) of six number of mirrors as input parameters for CST. In addition, the parameters like time (t) in minute, global solar radiation (G) in W/m^2, and temperature (T) in °C are considered as input parameter for optimizing the first law and second law efficiency and the results are depicted in Table 3.

During the optimization, first, data generation of the CST is done via experimentation under different input conditions and the experimental results are tabulated in Table 1, Table 2, and Table 3, respectively. First, data of Table 1, Table 2, and Table 3 are taken as input decision matrix for the optimization. Second, determination of weights for each of the CST input parameters is done using Equations (4-5) (Suman et al., 2019 & Nixon, 2013), and results are tabulated in Table 4, Table 5, and Table 6. After that formulation of weighted decision matrix and determination of MPRI values for each of the CST output parameter is done using Equations (6-8) (Suman et al., 2019 & Nixon, 2013) and the results are as tabulated in Table 4, Table 5, and Table 6.

RESULTS AND DISCUSSIONS

Based on the detailed modeling of the considered CST system (thermal modeling and the proposed MCDM modeling), parametric optimization has been done for improving the performance of the same. A detailed parametric analysis is done using an experimental approach to elucidate how the different input parameters affect the output parameters as detailed in Table 1 and Table 2.

Parametric Analysis

Performance of CST system depends on its design parameters. Various design parameters like altitude angle (α), length (L), heliostat tilt angle (β) affect the performance of CST. The height of the receiver and outlet temperature (T) of the heat transfer fluid i.e. the working fluid directly affect the efficiency of the CST. Improper selection of any of these design parameters leads to reduction in outlet temperature (T), as well as reduction in the performance of the CST. The ultimate performance of the CST is evaluated in terms of its first law efficiency or the thermal efficiency, and the second law efficiency or the exergetic efficiency. In the parametric analysis, first the effects of summer altitude angle and winter altitude angle on the summer height and winter height of the central receiver are analysed in Figure 6. The variations are plotted with respect to trial number. Ten different trial numbers are considered at different timings during the day. It can be seen that summer altitude angle and summer heights are greater than winter altitude angle and winter heights. This is attributed to higher declination angle of summer season compared to winter season. Further summer height and winter height of the receiver decrease with the passage of time during the day whereas summer altitude angle and winter altitude angle increase with the passage of time. This is attributable to the variation of Sun's azimuth angle as the Sun traverses from East to West.

The variations of altitude angle and the receiver height impact the temperature of the working fluid of the receiver. In order to investigate the composite effect of mirror tilt angle and solar irradiation on the maximum attainable temperature of the working fluid in the receiver (T), Figure 7 is plotted. Figure 7 is plotted for the 3rd set of readings as given in Table 2. As can be seen from the plot, the solar radiations are reflected from six set of mirrors onto the receiver unit. The maximum temperature of the working fluid increases to 51°C from 27°C considered as initial temperature of the

Figure 6. Plot of altitude angle Summer, altitude angle Winter on summer receiver (hrs) height and winter height of the receiver

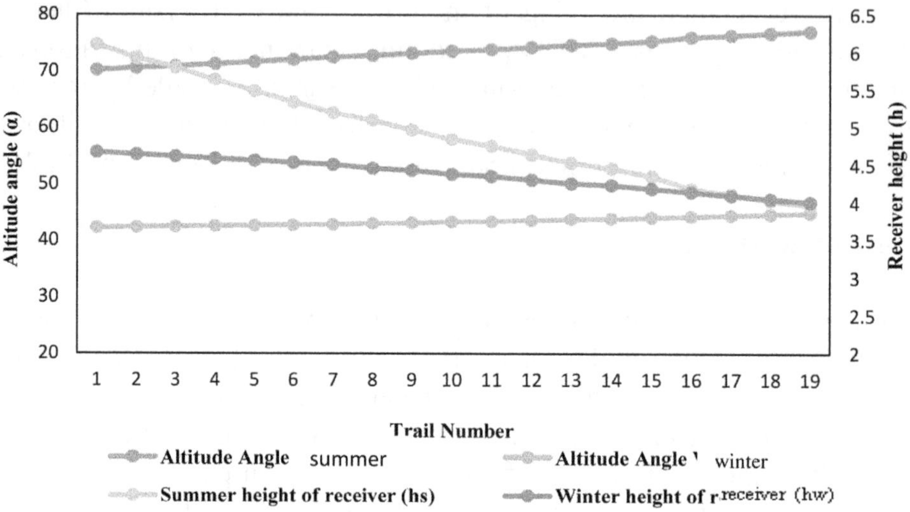

working fluid. With the corresponding increase of working fluid temperature, the tilt angles of all the mirrors are also increased in the range 30° to 65°. The variation of solar radiation follows normal weather cycle of the day of experimentation during with it has varied from a maximum 862 W/m^2 to a minimum 200 W/m^2 till the late afternoon. For the given arrangements of mirrors in terms of tilt angles and the receiver heights, the ambient and the operating parameters i.e. the global solar radiation and the working fluid temperature shall affect the ultimate performance of the CST system. Table 3 presents the results of thermal and exergetic efficiencies with respect to the obtained working fluid temperatures and the global solar radiations. These readings are corresponding to different times of a day i.e. like previous, three sets of readings (each set comprising of ten trial numbers) have been generated. The initial working fluid temperature is taken as 27°C, which is at the ambient condition. As the day progresses, the working fluid temperature increases till the afternoon time as the intensity of radiation increases. In each set of readings, the temperature increases by about 20°C on an average.

It can be observed from Table 3 that the thermal efficiency of the CST system is obtained in the range 0.112 to 0.827 and the exergetic efficiency is obtained in the range 0.013 to 0.072 combining all sets of readings.

Table 3. First law and second law efficiency for different temperature readings and global solar radiations

Trail No.	Time (t)	Global solar radiation (G) in W/m²	Temperature (T) in °C	Thermal efficiency (η)	Exergetic efficiency (η_e)
Set 1					
1	11:15:00	828.33	27	0.8004	0.0169
2	11:30:00	828.821	31	0.7023	0.0221
3	11:45:00	810.657	34	0.5832	0.0222
4	12:00:00	806.379	36	0.6180	0.0297
5	12:15:00	778.724	39	0.3545	0.0280
6	12:30:00	768.954	40	0.4843	0.0182
7	12:45:00	752.867	42	0.5399	0.0364
8	1:00:00	712.318	45	0.5839	0.0200
9	1:15:00	666.563	46	0.5281	0.0381
10	1:30:00	640.563	45	0.5839	0.0200
Set 2					
1	11:15:00	205.384	27	0.7649	0.0109
2	11:30:00	205.393	29	0.7024	0.0148
3	11:45:00	206.557	31	0.5296	0.0130
4	12:00:00	205.534	32	0.6837	0.0238
5	12:15:00	205.434	35	0.6332	0.0368
6	12:30:00	206.465	38	0.6369	0.0283
7	12:45:00	207.039	42	0.4583	0.0294
8	1:00:00	206.789	44	0.4616	0.0220
9	1:15:00	207.794	41	0.4417	0.0213
10	1:30:00	208.102	40	0.4389	0.0359
Set 3					
1	11:50:00	862.877	27	0.8276	0.0452
2	12:05:00	840.519	41	0.7146	0.0592
3	12:20:00	825.324	50	0.6591	0.0723
4	12:35:00	810.296	59	0.5676	0.0376
5	12:50:00	785.048	66	0.2781	0.0735
6	1:05:00	768.954	68	0.4074	0.0594
7	1:20:00	752.867	72	0.4352	0.0692
8	1:35:00	712.318	77	0.3672	0.0622
9	1:50:00	666.563	81	0.1120	0.0319
10	2:10:00	585.034	76	0.1391	0.0221

Figure 7. Plot of global solar radiation (G) and different title angles (β) of the mirrors on temperature (T) of CST system

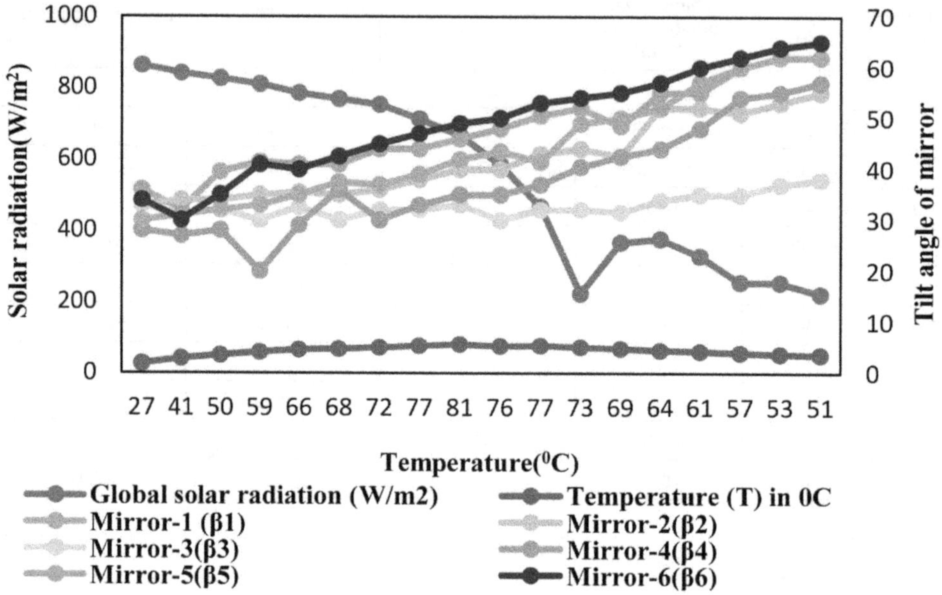

Figure 7 and Figure 8 show the variations of first law thermal efficiency and exergetic efficiency of the CST system with respect to the change of reduced temperature difference for set 1 and set 2 readings respectively. As can be seen from these figures, the trends in the variations of thermal efficiency and exergetic efficiency are the same as in case of modeling of solar flat plate collector, as discussed in chapter 2. With the increase of reduced temperature difference, the first law thermal efficiency decreases whereas the exergetic efficiency increases. As also explained earlier, the reason for this different behaviour in case of exergetic efficiency is the result of lowering of solar radiation and increase of inlet temperature to the receiver unit. The temperature reading for each observation under a given set is the inlet water temperature, which is also the outlet water temperature for the previous observation. Obviously with increase of working fluid temperature, the heat losses from the receiver increase and thus lower the thermal efficiency.

With regards the set 3 observations, the exergetic efficiency is found to decrease with the increase of reduced temperature difference unlike the previous two set of readings (Figure 8). This is the result of the increase of outlet temperature to higher values than the previous two cases as the

Figure 8a. Standard exergetic efficiency plot (set 1)

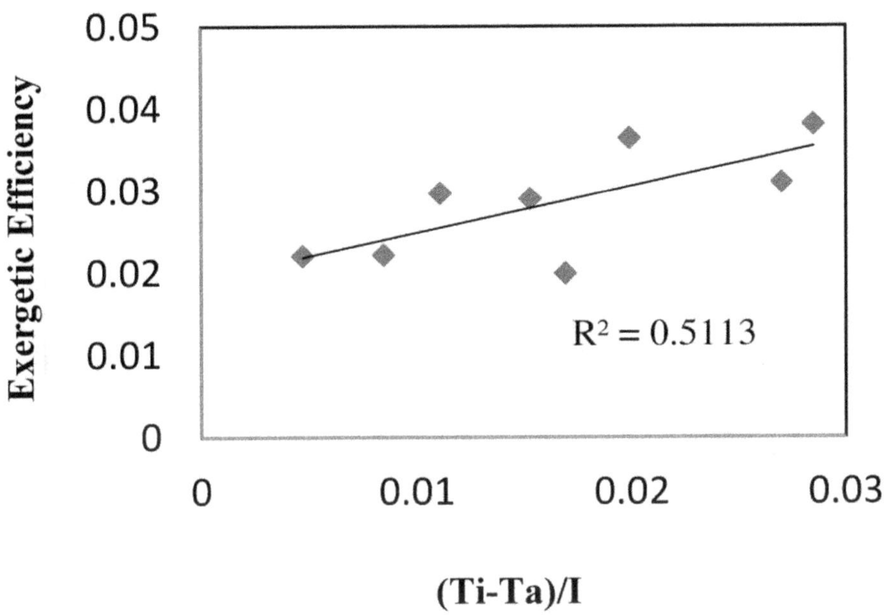

corresponding inlet temperatures are also higher. Thus, two high working fluid temperatures shall reduce the work potential of the heating process thereby lowering the exergetic efficiency even though solar radiation may reduce the heating effect. That is to say that exergetic efficiency of the CST can increase with the increase of reduced temperature difference whenever the working fluid temperature is increased upto a certain extent only. However, the thermal efficiency variation maintains the previous trend (Figure 8). Further it can be seen from these figures that the degree of correlation (R-squared value) for exergetic efficiency of set 3 is lower than the previous two sets as the trend deviate here; on the contrary the degree of correlation (R-squared value) for first law thermal efficiency of set 3 is better than the previous two sets.

Optimization of CST Parameters

As seen from the last section, there are so many design and operating parameters that control performance of the CST system, which are also conflicting in nature. Hence, this is ideally suited for decision making and optimization

Figure 8b. Standard thermal efficiency plot (set 1)

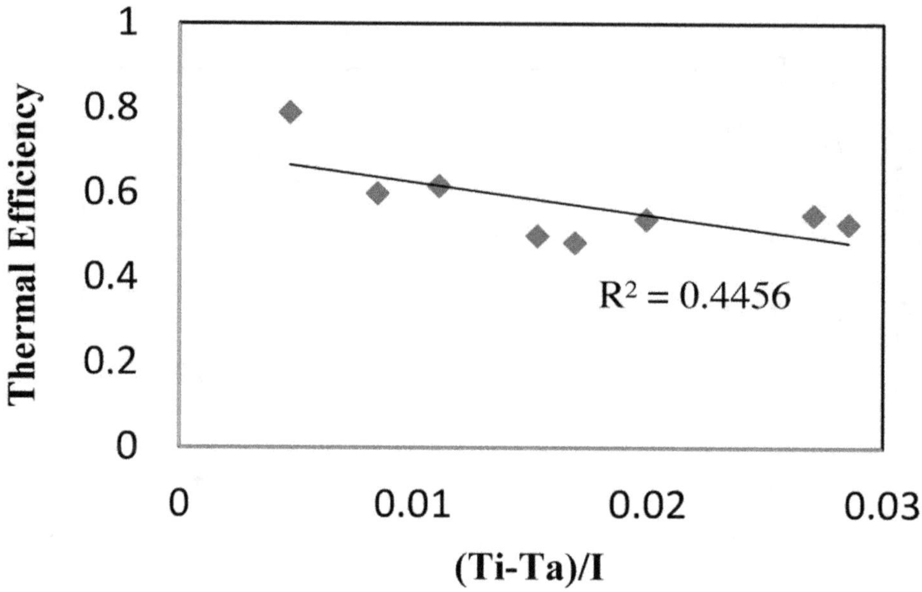

Figure 8c. Standard exergetic efficiency plot (set 2)

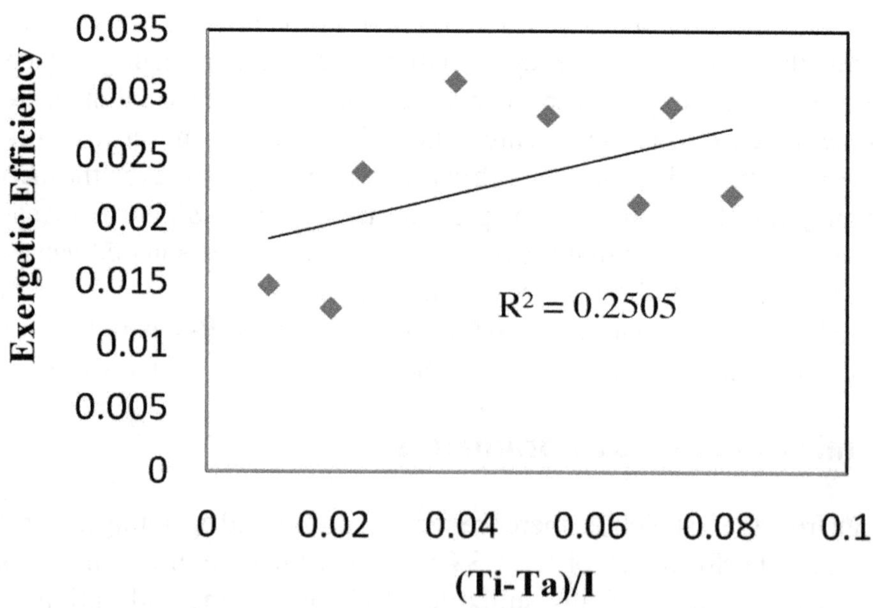

Figure 8d. Standard thermal efficiency plot (set 2)

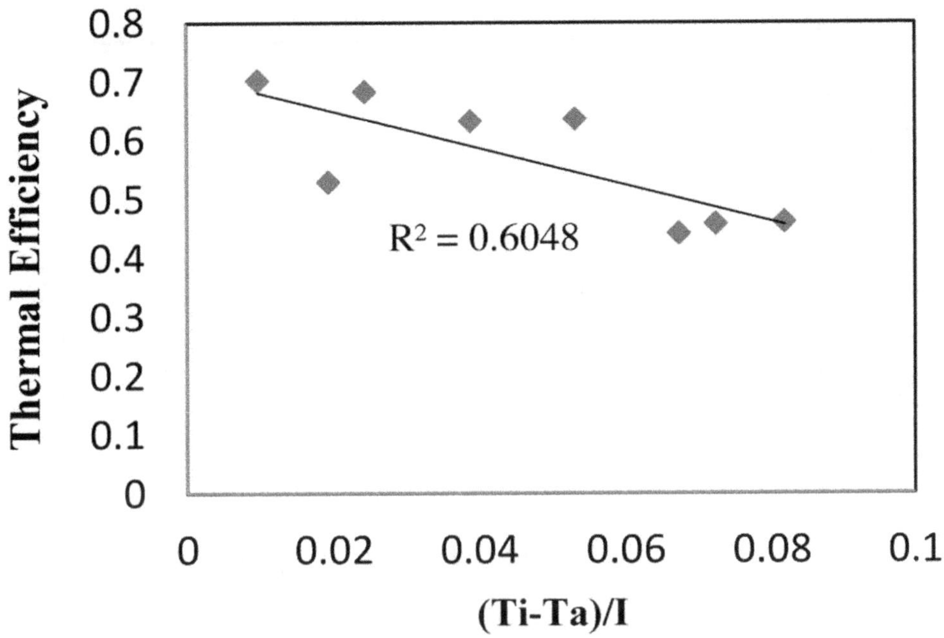

Figure 8e. Standard exergetic efficiency plot (set 3)

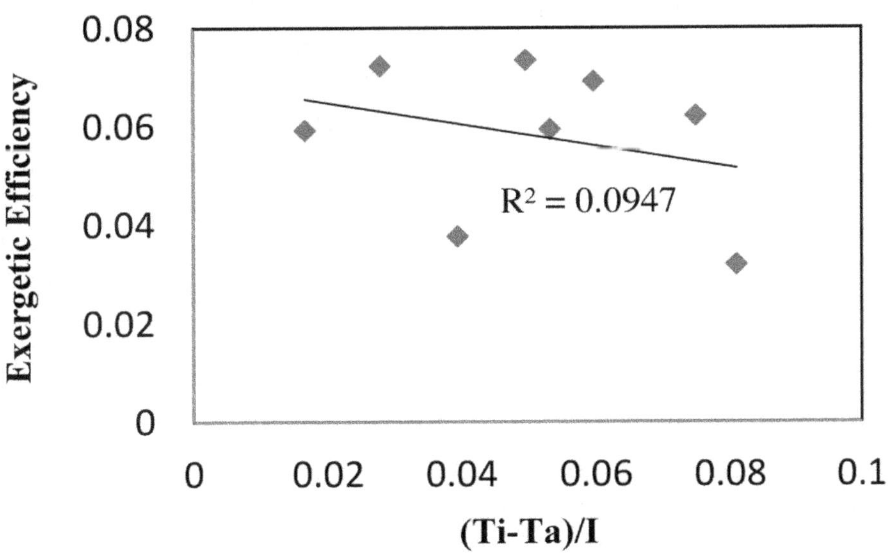

Figure 8f. Standard thermal efficiency plot (set 3)

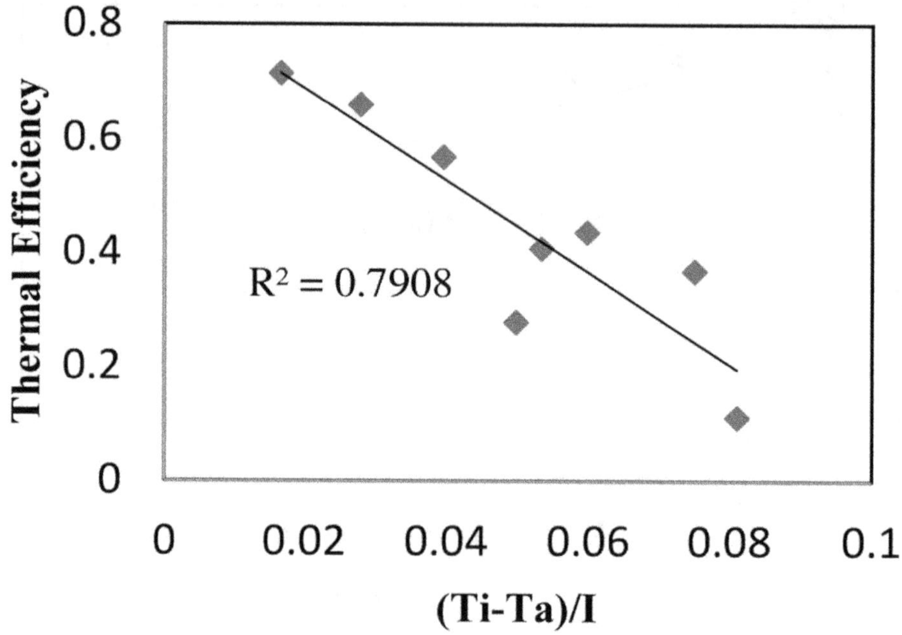

through MCDM techniques as discussed in this section. Optimization of CST is done based on MPRI values obtained in previous step. The DEAR method (Cavallaro, 2009) is used for modeling and optimization of CST parameters.

The results reveal that trail no-1 yields highest MPRI value for winter and summer season while Trail No. 9 for set-1 data; Trail No-8 for set-2 and Trail No-9 for set-3 yield highest MPRI value. The MPRI values for winter and summer season are obtained as 0.6618, and MPRI values as 1.1340 for set-1; 1.0375 for set-2 and 3.5161 for set-3 are obtained for different mirror angle and global solar radiation. The optimal setting for winter season is obtained as α = 42.2°, L=1.5 m and β= 32.8° and the optimal setting for summer season is obtained as α = 70.2°, L=1.5 m and β= 32.8°. Similarly, optimal settings for higher temperature are for set-1: [t=1:15 PM, G=666.563 W/m^2, β_1=41°, β_2=40°, β_3=36°, β_4=34°, β_5=37.5°, β_6=42.5°], for set-2 [t=1:00 PM, G=206.789 W/m^2, β_1=41°, β_2=39°, β_3=34.5°, β_4=31°, β_5=35°, β_6=39°], and for Set-3 [t=1:50 PM, G=666.563 W/m^2, β_1=46°, β_2=40°, β_3=33°, β_4=35°, β_5=42°, β_6=49°]. It is also observed that in general for global solar radiation (G) » 600-670 W/m^2 and tilt angles of β_1 » 41° to 46°, β_2 » 39°, β_3 » 34° to 36°, β_4 » 31° to 35°, β_5 » 35° to 42° and β_6 » 39° to 49° of the CST provide

Table 4. Weighted decision matrix for h_w and h_s and MPRI-I values of CST system

Trail No.	Weighted decision matrix for winter height of the receiver (h_w)	Weighted decision matrix for summer height of the receiver(h_s)	MPRI-I values
1	0.2628	0.399	0.6618
2	0.2597	0.376	0.6357
3	0.2567	0.3599	0.6166
4	0.2536	0.34	0.5935
5	0.2506	0.3217	0.5722
6	0.2477	0.3046	0.5524
7	0.2448	0.289	0.5338
8	0.2392	0.278	0.5171
9	0.2365	0.2641	0.5006
10	0.2311	0.2513	0.4824
11	0.2285	0.2421	0.4706
12	0.2233	0.2308	0.4541
13	0.2184	0.2201	0.4385
14	0.216	0.2125	0.4286
15	0.2113	0.203	0.4143
16	0.2068	0.1878	0.3945
17	0.2023	0.1817	0.384
18	0.198	0.174	0.372
19	0.1938	0.1686	0.3624

optimal temperature results. Subsequently, Trail No. 9 for set-1, Trail No-8 for set-2 and Trail No-3 for set-3 yield highest MPRI value. The MPRI values for winter and summer season are obtained as 0.7371 for set-1, 0.6744 for set-2 and 0.8708 for set-3 for different time (t), global solar radiation (G) and temperature (T) conditions. Furthermore, optimal parameter obtained via DEAR method provides highest operating height of receiver (h_s & h_w) for both winter and summer season as well as maximum temperature for different global solar radiation (G) and different tilt angle (β) for mirrors, which directly or indirectly increase the efficiency and performance of the CST system.

Table 5. Weighted decision matrix for T and MPRI-II values of CST system

Trail No.	Weighted decision matrix for (T)	MPRI-II values	Rank
Set-1			
1	0.01447	0.3907	10
2	0.01661	0.5150	9
3	0.01822	0.6195	8
4	0.01929	0.6945	7
5	0.02090	0.8151	6
6	0.02144	0.8574	5
7	0.02251	0.9453	4
8	0.02412	1.0852	3
9	0.02465	1.1340	1
10	0.02412	1.0852	2
Set-2			
1	0.01447	0.3907	10
2	0.01554	0.4507	9
3	0.01661	0.5150	8
4	0.01715	0.5488	7
5	0.01876	0.6565	6
6	0.02036	0.7738	5
7	0.02251	0.9453	2
8	0.02358	1.0375	1
9	0.02197	0.9009	3
10	0.02144	0.8574	4
Set-3			
1	0.01447	0.3907	18
2	0.02197	0.9009	17
3	0.02680	1.3398	16
4	0.03162	1.8655	13
5	0.03537	2.3344	9
6	0.03644	2.4780	8
7	0.03859	2.7781	6
8	0.04126	3.1774	2
9	0.04341	3.5161	1
10	0.04073	3.0954	4
11	0.04126	3.1774	3
12	0.03912	2.8558	5
13	0.03698	2.5514	7
14	0.03430	2.1951	10
15	0.03269	1.9941	11
16	0.03055	1.7412	12
17	0.02840	1.5054	14
18	0.02733	1.3939	15

Table 6. Weighted decision matrix for (η) and (η_e) and MPRI-III values of CST system

Trail No.	Weighted decision matrix for (η)	Weighted decision matrix for (η_e)	MPRI-III values	Rank
		Set-1		
1	0.0094	0.0136	0.2539	10
2	0.0108	0.0157	0.3348	9
3	0.0118	0.0172	0.4027	8
4	0.0125	0.0182	0.4514	7
5	0.0136	0.0197	0.5298	6
6	0.0139	0.0202	0.5573	5
7	0.0146	0.0212	0.6145	4
8	0.0157	0.0227	0.7054	3
9	0.0160	0.0232	0.7371	1
10	0.0157	0.0227	0.7054	2
		Set-2		
1	0.0109	0.0157	0.2539	10
2	0.0117	0.0169	0.2930	9
3	0.0125	0.0181	0.3348	8
4	0.0129	0.0186	0.3567	7
5	0.0141	0.0204	0.4267	6
6	0.0153	0.0221	0.5030	5
7	0.0169	0.0245	0.6145	2
8	0.0177	0.0256	0.6744	1
9	0.0165	0.0239	0.5856	3
10	0.0161	0.0233	0.5573	4
		Set-3		
1	0.0101	0.0147	0.2539	7
2	0.0154	0.0223	0.5856	4
3	0.0188	0.0272	0.8708	1
4	0.0221	0.0321	0.2126	8
5	0.0248	0.0359	0.5174	6
6	0.0255	0.0370	0.6107	3
7	0.0270	0.0392	0.8058	2
8	0.0289	0.0419	0.0653	9
9	0.0304	0.0441	0.2855	5
10	0.0285	0.0413	0.0120	10

Analysis of Variance Analysis

Further, Analysis of Variance (ANOVA) analysis with 95% of Confidence Interval is used to determine the significance and adequacy of the developed model. In the ANOVA analysis, the Fisher's statistical test (F-Ratio) and significance probability (P-value) followed by coefficient of determination (R^2) and adjusted-R^2 are determined to test the sufficiency and fitness of the proposed model. The F-value denotes the most influencing parameters on the performance characteristics while P-value shows the statistical significance of process variables of CST system. During the analysis, process variables which have larger F-values and lower P-values ($P<0.05$) indicate their significance statistically (Aragonés-Beltrán, et al. 2014). The R^2 signifies the total variation in the response expected by the model. The purpose of adjusted-R^2 is to analyse the model adequacy and fitness. In this research stepwise elimination method is used to remove the insignificant parameters to adjust the fitted quadratic model (Muthuramalingam, et al., 2018). The ANOVA analysis results are tabulated in Tables 7-9.

Table 7. ANOVA for MPRI-I values of CST system

Source	DF	Adj SS	Adj MS	F-Value	P-Value
Regression	2	0.15504	0.07752	21711087	0
Summer height of the receiver (h_s)	1	0.000386	0.000386	108172.8	0
Winter height of the receiver (h_w)	1	0.004068	0.004068	1139286	0
Error	16	0	0		
Total	18	0.15504			
R^2=99.97%, R^2(adj)=99.14%, R^2 (Pred)= 98.77%					

The ANOVA for MPRI-I results show (Table 7) that the parameter summer height of the receiver (h_s) and winter height of the receiver (h_w) are the most significant factors influencing the assessment of MPRI-I values of CST performance. Whereas Altitude Angle Summer (α), Length of mirror, Different title angle β (deg.) Altitude Angle Winter (α) are considered as statistically insignificant since they failed to test their significance in the ANOVA analysis. Subsequently, the P values for summer height of the receiver (h_s) and winter height of the receiver (h_w) are found to be less than 0.05, which indicate that the model and associated terms are statistical significant (Manoj et al., 2018). The other parameters except parameter highlighted in Table 7 are considered to be statistically insignificant. The values of R^2 are calculated to be 0.9997 for MPRI-I values, which imply that 95% of experimental data are well-suited. Higher R^2 coefficient indicates the satisfactory model to the experimental data. Finally, the adequacy and fitness of the model is calculated by adjusted-R^2 values. The high value of adjusted-R^2 (0.9977 for MPRI-I values) supports a high correlation between the experimental and the predicted values (Kulekci, et al., 2011).

Table 8. ANOVA for MPRI-II values of CST system

Source	DF	Adj SS	Adj MS	F-Value	P-Value
Regression	1	32.323	32.323	2085.44	0
Temperature (T)	1	32.323	32.323	2085.44	0
Error	36	0.558	0.0155		
Lack-of-Fit	28	0.558	0.0199		
Pure Error	8	0	0		
Total	37	32.881			
R^2=98.30%, R^2(adj)=98.24%, R^2 (Pred)= 98.04%					

The ANOVA for MPRI-II result shows (Table 8) that the parameter summer temperature (T) is the most significant factor influencing the assessment of MPRI-II values of CST performance. Subsequently, the P values for summer temperature (T) are found to be less than 0.05, which indicate that the model and associated terms are statistically significant (Manoj et al., 2018). The other parameters except parameter highlighted in Table 8 are considered to be statistically insignificant. The value of R^2 is calculated as 0.9830 for MPRI-II values, which implies that 95% of experimental data are well-suited. Higher R^2 coefficient indicates the satisfactory model to the experimental data. Finally, the adequacy and fitness of the model is calculated by adjusted-R^2 values. The high value of adjusted-R^2 (0.9804 for MPRI-II values) supports a high correlation between the experimental and the predicted values (Kulekci, et al.,2011).

The ANOVA for MPRI-III result shows (Table 9) that the parameters both thermal efficiency (η_t) and exergetic efficiency (η_e) are most significant

Table 9. ANOVA for MPRI-III values of CST system

Source	DF	Adj SS	Adj MS	F-Value	P-Value
Regression	2	0.01334	0.006672	0.14	0.001
Thermal Efficiency (η_t)	1	0.0082	0.008205	0.17	0.003
Exergetic Efficiency (η_e)	1	0.00828	0.008283	0.17	0.001
Error	27	1.29585	0.047995		
Lack-of-Fit	26	1.29585	0.049841		
Pure Error	1	0	0		
R^2=98.78%, R^2(adj)=97.98%, R^2 (Pred)= 97.25%					

factors influencing the assessment of MPRI-III values of CST performance. Subsequently, the P values of thermal efficiency (η_t), and exegetic efficiency (η_e) are found to be less than 0.05, which indicate that the model and associated terms are statistically significant (Manoj et al., 2018). The other parameters except parameter highlighted in Table 8 are considered to be statistically insignificant. The value of R^2 is calculated to be 0.9878 for MPRI-III values, which implies that 95% of experimental data are well-suited. Higher R^2 coefficient gives the satisfactory model to the experimental data. Finally, the adequacy and fitness of the model is calculated by adjusted-R^2 values. The high value of adjusted-R^2 (0.9725 for MPRI-III values) supports a high correlation between the experimental and the predicted values (Kulekci, et al.,2011).

Confirmatory Test

Additionally, confirmatory tests are conducted to verify or validate the optimal setting obtained via the proposed method. The optimal setting for higher temperature i.e. for set-1: [t=1:15 PM, G=666.563 W/m^2, β_1=41°, β_2=40°, β_3=36°, β_4=34°, β_5=37.5°, β_6=42.5°] for set-2 [t=1:00 PM, G=206.789 W/m^2, β_1=41°, β_2=39°, β_3=34.5°, β_4=31°, β_5=35°, β_6=39°], and for set-3 [t=1:50 PM, G=666.563 W/m^2, β_1=46°, β_2=40°, β_3=33°, β_4=35°, β_5=42°, β_6=49°] are used for confirmatory experiments and the corresponding results are shown in Table. 3.10. The results show that the confirmatory tests results are comparable and acceptable with experimental results for the optimal conditions.

Table 10. Confirmatory tests for CST thermal system

Input Conditions	Output Parameters	Experimental results	Confirmatory results
Set-1: [t=1:15 PM, G=666.563 W/m^2, β_1=41°, β_2=40°, β_3=36°, β_4=34°, β_5=37.5°, β_6=42.5°	Temperature (T) in °C	46	45
Set-2: [t=1:00 PM, G=206.789 W/m^2, β_1=41°, β_2=39°, β_3=34.5°, β_4=31°, β_5=35°, β_6=39°] and	Temperature (T) in °C	44	41
Set-3: [t=1:50 PM, G=666.563 W/m^2, β_1=46°, β_2=40°, β_3=33°, β_4=35°, β_5=42°, β_6=49°]	Temperature (T) in °C	41	42

SUMMARY

Modeling and optimization of CST using DEAR method is discussed. The present work optimized the CST parameters under two conditions. First- determination of optimal height of the receiver (*h*) for summer and winter conditions of CST. Second, the determination of optimal temperature (*T*) for different global solar radiation (*G*) and different title angle (β) of six (6 nos.) number of mirrors. Further, parametric analysis is explored to study the effect of individual parameter on CST performance.

From the study it is concluded that performance of CST under considered environments is feasible and use of DEAR method provides optimal solution. The results reveal that optimal setting for winter season is α = 42.2°, *L*=1.5 m and β= 32.8° and the optimal setting for summer season is α = 70.2°, L=1.5 m and β= 32.8°. Also the optimal setting as global solar radiation (*G*) » 600-670 W/m^2, tilt angle β_1 » 41° to 46°, β_2 » 39°, β_3 » 34° to 36°, β_4 » 31° to 35°, β_5 » 35° to 42° and β_6 » 39° to 49° for highest temperature are obtained. Additionally, ANOVA analysis is carried out to determine the significance and adequacy of the developed model. The results show that all the response parameters are the major parameters for CST system and also model fitted data are well fitted with experimental result within 95% confidence level.

Furthermore, optimal parameter obtained via DEAR method provides highest operating height of receiver (h_s & h_w) for both winter and summer season as well as maximum temperature for different global solar radiation (*G*) and different tilt angle (β) for the mirrors, which directly or indirectly increase the efficiency and performance of the CST system. At last, confirmatory tests show comparable and acceptable results with experimental values. Finally, as the MCDM technique used give optimized values of the considered design and operating parameters and efficiency of the collector, and as these are obtained with good accuracy and with statistical significance, the obtained optimized values can be practically implemented for the present solar collector.

REFERENCES

Aragonés-Beltrán, P., Chaparro-González, F., Pastor-Ferrando, J. P., & Pla-Rubio, A. (2014). An AHP (Analytic Hierarchy Process)/ANP (Analytic Network Process)-based multi-criteria decision approach for the selection of solar-thermal power plant investment projects. *Energy*, *66*, 222–238. doi:10.1016/j.energy.2013.12.016

Cavallaro, F. (2009). Multi-criteria decision aid to assess concentrated solar thermal technologies. *Renewable Energy*, *34*(7), 1678–1685. doi:10.1016/j.renene.2008.12.034

Desai, N. B., & Bandyopadhyay, S. (2017). Line-focusing concentrating solar collector-based power plants: A review. *Clean Technologies and Environmental Policy*, *19*(1), 9–35. doi:10.100710098-016-1238-4

Diemuodeke, E. O., Addo, A., Oko, C. O. C., Mulugetta, Y., & Ojapah, M. M. (2018). Optimal Mapping of Hybrid Renewable Energy Systems for Locations Using Multi-Criteria Decision-Making Algorithm. *Renewable Energy*, 1–6.

Kadek, A., Riyanarto, W., & Nurul, F. A. (2016). Optimization Solar Farm Site Selection Using Multi Criteria Decision Making Fuzzy AHP and PROMETHEE: Case Study in Bali. *International Conference on Information, Communication Technology and System (ICTS)*, 237-243.

Kulekci, M.K., Esme, O. E., & Kazancoglu, Y. (2011). Modeling and prediction of weld shear strength in friction stir spot welding using design of experiments and neural network. *Materials science and Engineering Technology*, *42*, 990-995.

Manoj, M., Jinu, G. R., & Muthuramalingam, T. (2018). Multi response optimization of AWJM process parameters on machining TiB2 particles reinforced Al7075 composite using Taguchi-DEAR methodology. *Silicon*, *10*(5), 2287–2293. doi:10.100712633-018-9763-x

Mehrabi, D., Mohammadzaheri, M., Firoozfar, A., & Emadi, M. (2017). A fuzzy virtual temperature sensor for an irradiative enclosure. *Journal of Mechanical Science and Technology*, *31*(10), 4989–4994. doi:10.100712206-017-0947-x

Mevlut, U. (2013). GIS-based solar farms site selection using Analytic Hierarchy Process (AHP) in Karapinar region, Konya/Turkey. *Renewable & Sustainable Energy Reviews*, *28*, 11–17. doi:10.1016/j.rser.2013.07.042

Mohanty, S., Rout, A., Patra, P. K., & Sahoo, S. S. (2017). ANFIS based solar radiation data forecasting for energy and economic Study of solar water heaters in eastern India. *International Journal of Control Theory and Applications*, *10*, 179–190.

Muthuramalingam, T., Vasanth, S., Vinothkumar, P., Geethapriyam, T., & Rabik, M. M. (2018). Multi criteria decision making of abrasive flow oriented process parameters in abrasive water jet machining using Taguchi-DEAR methodology. *Silicon*, *10*(5), 2015–2020. doi:10.100712633-017-9715-x

Nixon, J. D., Dey, P. K., & Davies, P. A. (2013). Design of a novel solar thermal collector using a multi-criteria decision-making methodology. *Journal of Cleaner Production*, *59*, 150–159. doi:10.1016/j.jclepro.2013.06.027

Purohit, I., Purohit, P., & Shekhar, S. (2013). Evaluating the potential of concentrating solar power generation in Northwestern India. *Energy Policy*, *62*, 157–175. doi:10.1016/j.enpol.2013.06.069

Romero, M., & Augilar, J. G. (2013). Solar thermal CSP technology. *WIRES Energy and Environment*, *3*(1), 42–59. doi:10.1002/wene.79

Suman, D., Jagannath, R., & Biplab, D. (2019). Investigation of thermal performance of SAC using Fuzzy logic-based expert system. *Journal of Mechanical Science and Technology*, *33*(8), 4013–4021. doi:10.100712206-019-0543-3

Ummadisingu, A., & Soni, M. (2011). Concentrating solar power–technology, potential and policy in India. *Renewable & Sustainable Energy Reviews*, *15*(9), 5169–5175. doi:10.1016/j.rser.2011.07.040

Chapter 4
Modeling and Optimization of Evacuated Tubular Solar Thermal Collector

ABSTRACT

Modeling and optimization of evacuated tubular solar thermal collector (ETSTC) is discussed using a modified simple additive weighting (M-SAW) method. To improve the system efficiency (η) and end day temperature (Tsfd), ETSTC parameter (i.e., start day temperature [Tsid], ambient temperature [Tad], global solar radiation on tilted surface [GT], and wind speed [Ws]) are optimized. The applied method is significantly improved the efficiency (η) and determined the best setting for ETSCT. Test no.10 is the optimal experimental trail run and corresponding collector efficiency is obtained as 43%. Further, experimental data are statistically tested via parametric, ANOVA analysis, and found satisfactory and acceptable. Last, confirmatory tests results show comparable and acceptable w.r.t. experimental results for the optimal setting obtained through proposed method. The proposed MCDM method can be recognized as potential use for modeling and optimization of other thermal systems.

DOI: 10.4018/978-1-7998-3523-3.ch004

INTRODUCTION

The solar water heaters using solar thermal collectors are primarily the energy harvesters used worldwide, and the evacuated-tube designs of the collector are the most popular due to their simplicity and best overall performance over their counterpart flat-plate collectors especially in adverse weather conditions. Like flat plate collectors, evacuated tube collector utilizes both direct and diffuse part of the solar radiation. The evacuated tubular collector is more efficient than flat plate collector as the heating time required by this collector is less than its counterpart flat plate collector and also due to evacuation of the space in the annulus thereby reducing convection and conduction heat losses. Thus, an evacuated tube collector can operate at a higher temperature than a flat plate collector. Further, it can provide the same heating effect in less time than the latter collector. It has higher efficiency at low incidence angles, which tends to give it an advantage over flat plate collector in day-long performance. And in cloudy weather, the performance of ETSTC is better than solar flat plate collector. Development in evacuated tube design is important among which the water-in-glass design is very popular because of its low cost and simple manufacturing and also simple installation procedures. One more design uses a heat-pipe system with an intermediate fluid used to carry the heat from the heating elements to the tank (Kalogirou, 2004). A water-in-glass tube-based collector consists of a set of glass tubes connected to a shell or tank. Each tube is surrounded by a second glass tube of a higher diameter. The annular space between the tubes is evacuated in order to reduce the convective heat losses. Another variant of this design is the use of parabolic mirror reflectors attached underneath each tube for reflecting and concentrating radiant flux onto the tubes. The working medium in the evacuated tubes, generally water (or any better heat transfer fluid, like glycol), flows from the tank to the tubes, captures heat, and then return back to the tank by a natural flow mechanism (Morrison et al., 2004). Several types of research have been conducted to illustrate the overall performance of water-in-glass evacuated tube collectors and the results show that the overall efficiency is in the range of 50- 60% (Morrison, et al., 2005; Kim, & Seo, 2007; Zhang, & Yamaguchi, 2008). Innovative mathematical methods have been also used to investigate the performance and to find possible ways to improve existing designs of evacuated tubes (Hayek, 2009). A schematic diagram of this stationery collector based water heating system is shown in Figure 1 (a). The collector is formed into the shape of a number of tubes through which water is passed. The tubes are of two types- all glass type for e.g. Figure 1(b), and a heat pipe (copper pipe) design hermetically

Figure 1a. Photograph of a standard evacuated

sealed in a vacuum-sealed borosilicate glass tube, as shown in Figure 1 (b).

A heat-pipe based evacuated tube collector involves generally a similar set of tubes connected to a tank. Each tube is made of a finned copper pipe (heat pipe) surrounded by a glass tube and the annular space in between is evacuated. Heat pipes are structures of very high thermal conductance. They permit the transport of heat with a temperature drop, which are several orders of magnitude smaller than for any solid conductor of the same size. The heat pipe is a closed container consisting of a capillary wick structure and a small amount of vaporizable fluid. It is based on an evaporating-condensing cycle involving an evaporation phase in which heat is transferred as latent heat energy to the working fluid of the tubes by using the solar heat, which is

Figure 1b. Schematic of an evacuated tube tubular collector (Courtesy MNRE website)

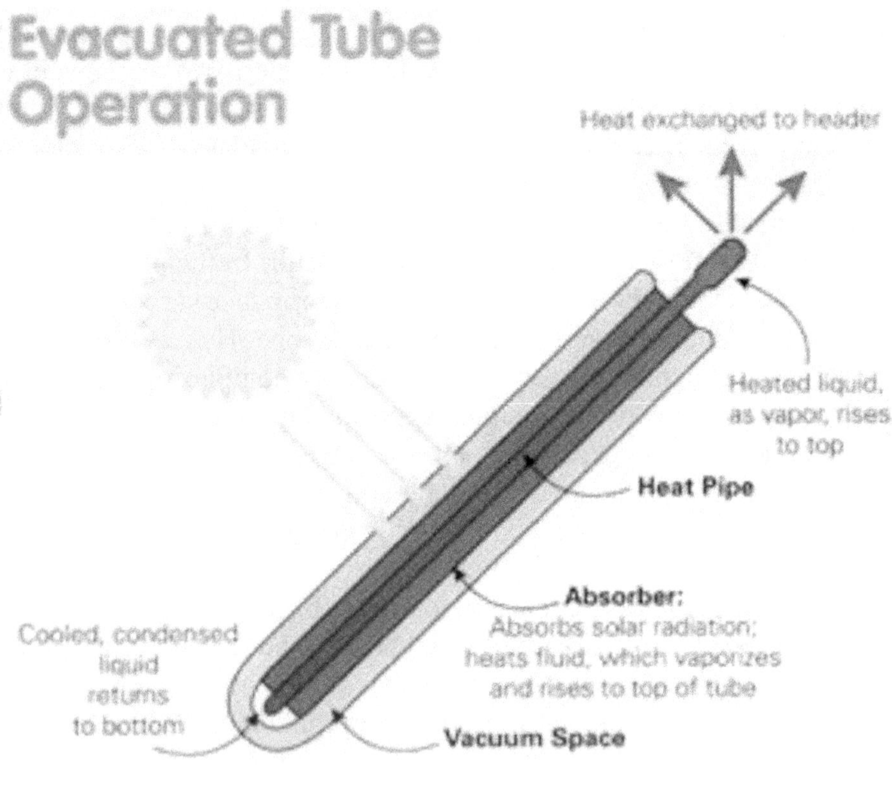

followed by a condensation phase in which that heat is released by the working fluid to water in a metal manifold or a heat sink (i.e. condenser) located at the top of the tube or at the junction of tank and the tubes. The circulation is completed by return ñow of the condensate to the heating zone. Thus, the working fluid flows by natural circulation between the two phases in order to transport the heat as needed (Walker, et al., 2004 & Kim, & Seo, 2007). If the working fluid is other than water, then the heat released in the heat sink is to be picked up by water through a heat exchanger across the heat sink. The major benefit of this design is the self-limiting temperature control due to the inherent protection from freezing and overheating, which is the result of no evaporation or condensation of the working fluid above the phase-change temperature. Evaluation of the overall performance of the solar collectors is usually carried out experimentally using proven procedures according to international standards (Ashrae, 2010; Hill, & Streed, 1976; Proctor, 1984) and many correlations have been developed for the purpose of predicting the overall efficiency under various climatic conditions. Various studies have been conducted to evaluate the thermal performance of evacuated-tube solar collectors and to compare them to their counterpart flat plate collector (Zambolin, & Del, 2010).

Furthermore, a performance study of evacuated tubular solar collector is also essential and various factors are dependent on its performance. The effect of too much variation of the input parameters results in an error in the performance estimation of the evacuated tubular solar collector. In the previous studies, conventional techniques for the selection of performance variables were done by using a database or using operators' experience. Hence, a performance study of evacuated tubular solar collector tends to complex optimization task and contains a degree of indecision, vague and ambiguity information from the data collected from experimental trails and experts end. Thus, modeling and optimization of Evacuated tube solar collector are necessary. Kalogirou, (2004) carried out a design technique to determine the best combination of the collector across section area and storage tank volume to increase the life cycle savings of solar installation by adopting hybrid artificial neural networks (ANN) with genetic algorithm (GA) method. Bornatico et al. (2014) proposed a technique for investigating the optimal size of the main components for a solar thermal system using particle swarm optimization (PSO). Mohanty et al. (2017) exhibited the application of adaptive neuro-fuzzy inference system (ANFIS) technique for energy and economic analysis of solar water heater. Mehrabi et al. (2017) proposed a fuzzy interference system based virtual sensor algorithm for optimum prediction of temperature control points within the enclosure for solar system. Wiguna et al. (2016) carried out a case study in Bali to determine the best site for constructing a solar farm using a hybrid MCDM technique, fuzzy-analytic hierarchy process (AHP) and preference ranking organization method for enrichment of evaluations (PROMETHEE). Mevlu, (2013) proposed GIS-based solar farms site selection using AHP in Karapinar region, Konya/Turkey. Hassan, et al., (2017) proposed Solar PV power plant site selection using a GIS-AHP based approach with an application in Saudi Arabia. Hassan, & Al, (207) proposed a Fuzzy AHP and GIS-based approach to prioritize utility-scale solar PV sites in Saudi Arabia. Jagannath et al. (2019) carried out a case study on the effect of barriers in green supply chain management using modified SAW technique. Suman et al., (2019) investigated the thermal performance of solar air collector using Fuzzy logic based expert system. After this literature review, it is observed that the above mentioned studies have ignored the considerations of criteria weights as well as have not accounted for beneficial and non-beneficial parameters, which result in error in the optimal ranking. Furthermore, most of the MCDM methods used for solar system have rank reversal problem, which ultimately alters the optimal settings. In order to resolve these issues, the present chapter employs modified-SAW method for modeling and optimization for ETSTC. Also, no literature on modeling and optimization of ETSCT using Modified SAW method is available.

Experimental Details

The ETSTC is tested according to the standards laid down by the Bureau of Indian Standard, IS 16544:2016 (Morrison, et al., 2004) (for leakage and stagnation test, and required ambient condition), and IS 16368:2015 (Morrison, et al., 2003) (for system performance test). According to these standards (Morrison, et al., 2004), the daily average ambient temperature must be greater than 8°C, and the daily solar irradiation received by the ETSTC must be greater than 17 MJ/(m^2-day). The various experimental test conditions for the testing ETSTC system as laid down by the Bureau of Indian standard

Table 1. Experimental test conditions for ETSTC system

Sl. No	Experimental parameters	Conditions
1.	Pressure	0.06 MPa
2.	Daily average ambient temperature for stagnation test	>8°C
3.	Daily solar irradiation	must be greater than 17 MJ/(m^2-day)
4.	Duration of test	7 hours (3.5 hrs before noon and 3.5 hours afternoon)
5.	Wind velocity	< 4m/s
All the tests are to be conducted at normal temperature twice on the sample.		

(IS 16544:2016) are given in Table 1.

The experiments are conducted at various times in a day. The specifications of the individual instruments used are as follows- for measuring global solar radiation, a solar pyranometer (range 0-1500 W/m^2, accuracy ± 1W/m^2) is placed in the aperture plane of the ETSTC; for measuring water i.e. fluid temperature, PT 100 RTD type temperature sensor is used (range 0 to 150°C, accuracy ± 0.5°C); for measuring flow rate, an electronic flow meter is used (range 0.5-25 lpm, accuracy ± 1%). The other details of the experimental setup and conditions are discussed in the subsequent section. Various test runs on ten different days are conducted. The different experimental readings are given in Table 2. In this table second to seventh column terms represent water temperature of the system at the start of the day-test, at the end of the day-test, ambient temperature (T_{ad}), during the test period, tilted global solar radiation, non-dimensional reduced temperature (X), system efficiency (η_{system}), and wind speed (W_s) respectively. The system efficiency is calculated by using the following relation (Morrison, et al., 2004):

$$\eta_{sys} = \frac{(MC)\times(T_{sfd} - T_{sid})}{A_C \int_{t_1}^{t_2} G_t dt} \tag{1}$$

where, T_{sfd}= final storage water temperature; T_{sid}= Initial value of storage water temperature; G_t = solar irradiation; A_c= surface area of the collector.

The calculation of X involves the integration of storage water temperature over the duration of the daytime test.

$$X = \frac{\int_{t_1}^{t_2} T_{sd} - T_{ad}}{\int_{t_1}^{t_2} G_t dt}$$

$$X = \frac{\sum_{m=1}^{42}\left(\frac{T_{sfm+T_{sim}}}{2} - T_{adm}\right)\times(t_{2m} - t_{1m})}{\sum_{m=1}^{42}\left(\frac{G_{Tfm-G_{Tim}}}{2}\right)\times(t_{2m-t_{1m}})} \tag{2}$$

where T_{adm} represents average ambient temperature during the mth time interval; T_{sfm} and T_{sim} represent the water temperature of the RTDs in the storage tank at the end (i.e. t_{2m}) and start of the mth time interval (i.e. t_{1m}), respectively. Similarly, G_{Tfm} and G_{Tim} is the solar irradiance value recorded at the end and at the start of the mth time interval (Morrison, et al., 2004).

Table 2. Experimental results of ETSTC

Test no.	Input Parameters					Output parameters	
	Start day temperature (T_{sid}) in °C	Ambient temperature (T_{ad}) in °C	Global solar radiation (G_T) in W/m²	Non-dimensional reduced temperature (X)	Wind speed W_s (m/s)	System Efficiency(η) in %	End day temperature (Tsfd) in °C
1	32.5	28.4	500.0	0.038	0.6	0.42	48.0
2	35.4	27.5	639.4	0.026	1.8	0.37	51.6
3	37.1	24.3	595.5	0.042	0.8	0.39	53.3
4	42.9	26.8	643.0	0.035	1.9	0.46	60.3
5	45.0	26.8	612.8	0.05	0.3	0.40	62.8
6	50.1	27.2	609.7	0.045	1.5	0.42	65.8
7	51.2	28.2	572.5	0.052	0.2	0.37	65.9
8	58	30.7	575.3	0.041	1.3	0.41	72.6
9	61.0	26.8	624.1	0.06	0.4	0.36	73.2
10	62.9	30.7	658.4	0.062	1.8	0.43	76.9

4.3. Modified Simple Additive Weight (M-SAW) method

In this section, a Modified Simple Additive Weighted (M-SAW) method is proposed. Mainly, SAW (Afshari, et al., 2010 & Abrams, et al., 2018) is the most frequently used method in MCDM technique. The main advantage of this technique is simply the rank of factors and sub-factors that are evaluated using a weighted sum of all attributes values. The rank reversal is the weakness of this method, which appears when the factors are added or removed in the decision matrix. The present study applies to reduce the rank reversal in the SAW method introducing preference index value. The schematic flow chart of modified SAW method is shown in Figure 2. The proposed method consists of three steps: design of decision matrix, normalization of the decision matrix, determination of relative index value and final ranking based on the relative index value (Abrams, et al., 2018).

Step1: Design of decision matrix

The modified SAW method starts with the design of the decision matrix. The decision matrix is developed based on the number of criteria and alternatives. It contains performance values of each criteria corresponding to the number of alternatives. The typical design of decision matrix is expressed as in the following Equation (Afshari, et al., 2010).

$$A_{ij} = \begin{pmatrix} a_{11} & \cdots & a_{1n} \\ \vdots & \ddots & \vdots \\ a_{m1} & \cdots & a_{mn} \end{pmatrix} \tag{3}$$

where, A_{ij} is the decision matrix, a_{11}, a_{12},.......a_{mn}, performance values of *j*th alternative with respect to the *i*th criteria, *m* represents the number of alternatives and *n* denotes the number of criteria's.

Step 2: Formulation of a normalized weighted matrix

After constructing the decision matrix, normalization of the decision matrix is carried out using Equation (2). In this process, different units of

Figure 2. Schematic flow chart of M-SAW method for ETSTC

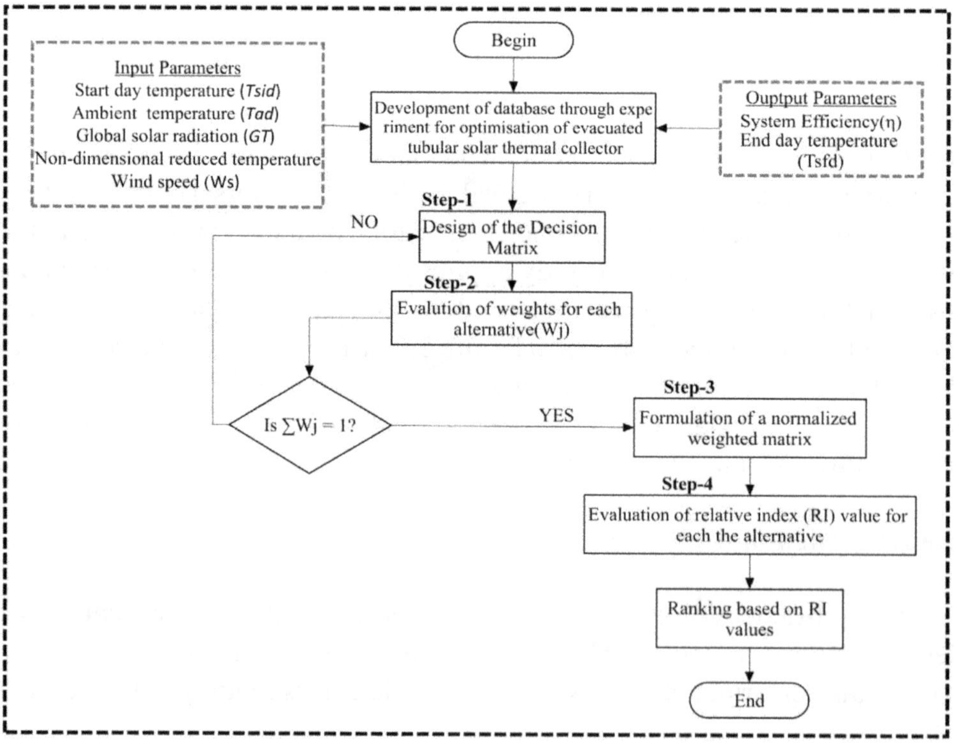

decision matrix values for each of the criteria's are converted into comparable sequence in the rage of 0 to 1. A normalized value for each of the jth criteria is determined as the ratio of performance value of *j*th criteria to the sum of column *j*th criteria's (Abrams, et al., 2018).

$$G_{ij} = \frac{A_{ij}}{\sum_{i=1}^{n} W_{ij}} \tag{4}$$

where, G_{ij} represents the normalized matrix with normalized performance values of *j*th criteria to the *i*th alternatives, $\sum_{i=1}^{n} W_{ij}$ is the sum of column total of the pair-wise matrix.

Next, the normalized weighted matrix is formulated by dividing the sum of the normalized matrix column of the matrix to the number of criteria's used (n) using Equation (5.) Based on the criteria weights the importance of criteria is decided i.e. criteria with higher weight shows much importance than criteria with lower weight (Afshari, et al., 2010 & Abrams, et al., 2018 & Kek, & Chin, 2018).

$$z_{ij} = \frac{\sum_{i=1}^{n} G_{ij}}{n} \tag{5}$$

where, z_{ij} is normalization weighted matrix, $\sum_{i=1}^{n} G_{ij}$ is the sum of rows of normalized decision matrix, *n* is number of criteria's.

Step 3: Relative index value and ranking

The Relative Index (RI) value for each the alternative is computed using Equation (6). This converts the multi-attribute problem into a single attribute problem with consideration of both beneficial and non-beneficial criteria's as well as the interrelationship between the criteria's into account (Abrams, et al., 2018).

$$R_{ij} = \frac{\sum_{i=1}^{n} G_{max} - G_i}{\sum_{i=1}^{n} G_i - G_{min}} \tag{6}$$

where, $\sum_{i=1}^{n} G_{max} - G_i$ represents the beneficial criteria, $\sum_{i=1}^{n} G_i - G_{min}$ is nonbeneficial criteria and R_{ij} denotes the relative index value. At last, the ranking of the alternatives is done based on the *RI*. The alternative with higher *RI* yields the optimal alternative while lower *RI* denotes the least optimal alternative for ETSTC.

Modeling of ETSTC System

The Modified SAW method (Afshari, et al., 2010 & Abrams, et al., 2018) is used for modeling and optimization for ETSTC parameters. The tests are conducted at different times in a day as well as at night. The various experiment runs on ten different days are conducted as shown in Table 2. In the present work, five independent parameters i.e. input parameters like start day temperature (T_{sid}) in °C, end day Temperature (T_{sfd}) in °C, ambient temperature (T_{ad}) in °C, global solar radiation (G_T) in W/m^2, non-dimensional

Table 3. Normalization of Decision matrix of ETSTC

Test no.	Normalized values (G_{ij}) for		Normalized Weighted (z) values
	System Efficiency (η)	End day temperature (T_{sfd})	
1	0.103	0.076	0.082
2	0.106	0.082	0.099
3	0.096	0.085	0.087
4	0.113	0.096	0.108
5	0.098	0.100	0.090
6	0.103	0.104	0.107
7	0.091	0.105	0.090
8	0.106	0.115	0.108
9	0.088	0.116	0.100
10	0.091	0.122	0.124

parameter (X) and two output parameters such as system efficiency (η) in % and end day temperature (T_{sfd}) in °C are considered. In modeling, first the design of a decision matrix is done using Equation (3), which contains a number of alternative as a number of trail nos. or input parameter settings and a number of criteria (Afshari, et al., 2010) as output parameters like system efficiency(η) in % and wind speed W_s in m/s for ETSTC. Second, the development of normalized decision matrix (Abrams, et al., 2018) and weighted normalized decision matrix are done by using the Equations. (4 & 5) respectively. The result of the normalized decision matrix (G_{ij}) and weighted normalized decision matrix (z_{ij}) are shown in Table 3.

Next, relative index values for each of the alternatives are determined using Equation (6). This step converts the multi-attribute problem into single attribute problem with consideration of both beneficial i.e. system efficiency (η) and end day temperature (T_{sfd}) as well as the interrelationship between the criteria's into account (Abrams, et al., 2018). At last, the ranking of the alternatives is done based on the *RI* values and ranking results are tabulated in Table 4. In this investigation, the lower weight of the parameter is considered first rank and higher weight is considered lower rank for the beneficial process, whereas non-beneficial parameters considered higher weight as the first rank and lower weight as the least rank as depicted in Table 4 and Table 5.

Table 4. Ranking of normalized weighted (z) values of ETSTC parameters (Beneficial)

Test no.	Start day temperature (T_{sid}) in °C	End day temperature (T_{sfd}) in °C	Ambient temperature (T_{ad}) in °C	global solar radiation (G_T) in w/m²	Parameter (X)	Normalized Weighted (z) values	Rank
1	32.5	48.0	28.4	500.0	0.038	0.329	10
2	35.4	51.6	27.5	639.4	0.026	0.209	6
3	37.1	53.3	24.3	595.5	0.042	0.290	9
4	42.9	60.3	26.8	643.0	0.035	0.144	3
5	45.0	62.8	26.8	612.8	0.05	0.273	8
6	50.1	65.8	27.2	609.7	0.045	0.150	4
7	51.2	65.9	28.2	572.5	0.052	0.270	7
8	58	72.6	30.7	575.3	0.041	0.141	2
9	61.0	73.2	26.8	624.1	0.06	0.200	5
10	62.9	76.9	30.7	658.4	0.062	0.031	1

Table 5. Ranking of normalized weighted (z) values of ETSTC parameters (Non-beneficial)

Test no.	Start day temperature (T_{sid}) in °C	End day temperature (T_{sfd}) in °C	Ambient temperature (T_{ad}) in °C	global solar radiation (G_T) in w/m²	Parameter (X)	Normalized Weighted (z) values	Rank
1	32.5	48.0	28.4	500.0	0.038	0.093	10
2	35.4	51.6	27.5	639.4	0.026	0.214	6
3	37.1	53.3	24.3	595.5	0.042	0.133	9
4	42.9	60.3	26.8	643.0	0.035	0.279	3
5	45.0	62.8	26.8	612.8	0.05	0.149	8
6	50.1	65.8	27.2	609.7	0.045	0.273	4
7	51.2	65.9	28.2	572.5	0.052	0.153	7
8	58	72.6	30.7	575.3	0.041	0.282	2
9	61.0	73.2	26.8	624.1	0.06	0.237	5
10	62.9	76.9	30.7	658.4	0.062	0.378	1

RESULTS AND DISCUSSION

Based on the thermal modeling and the proposed MCDM modeling of the ETSTC system, a parametric optimization has been done for improving the performance of the system. A detailed parametric analysis is done first using an experimental approach to understand how the different input parameters affect the output parameters as also given in Table 2. The experimental investigation of ETSTC system is conducted for the climatic condition of North eastern region of India located at 24.83° N, 92.77° E. The experimental datasets are generated at a technical university named National Institute of Technology Silchar, which is located in a city called Silchar in North-East of India. In general, climatic condition of Silchar is hot and humid during the month of April to September, the average temperature and humidity are 27°C and 86% respectively. The climatic condition of this region is characterised by frequent rainfall, high relative humidity, and the low annual average solar radiation (4.5 W/m²). The ETSTC experiments are conducted at various times in a day. Various test runs on ten different days are conducted. The performance of the present ETSTC system is evaluated in terms of the variation of its system efficiency and the useful heat gain over a time period equivalent of the heating period. Dynamic performance estimation (i.e. variation of performance with respect to time) of solar collector is required as the ambient parameters like solar radiation and ambient temperature are not constant over a given time period. Normally the heating period of solar thermal systems for e.g. water heaters is in between 9 a.m. to 3 p.m. In the present parametric study, the variations of all performance parameters are considered for an average solar insolation day and spanning over 10 such average days during which the daily solar irradiation received by the ETSTC is greater than 17 MJ/(m^2-day). In addition to that, the variations of ambient temperature and the intensity of solar radiation falling on the collector (also called solar irradiation) are monitored day-wise as they impact the performance of the evacuated collector. In addition, operating parameter like mass flow rate, water inlet temperature also impact the performance of the ETSTC. The following subsections are described the parametric analysis of the present study.

Parametric Analysis

Variation of Solar Radiation and Ambient Temperature

The variations of average solar radiation and ambient temperature over the considered 10 days period are shown in Figure 3(a). These two parameters are the major ambient parameters that control the performance of ETSTC. The solar radiation varies between 500 - 659W/m^2, whereas ambient temperature varies between 29°- 31°C. The fluctuations of all these values remain within narrow ranges, which will ensure almost uniform effects of these ambient parameters. And as per these ambient parameters, the variations of useful heat gain by the collector as well as its system efficiency are monitored during these many days, as shown in Figure 3b. Also Figure 3(c) shows the variation of the system efficiency (i.e. thermal efficiency) with respect to reduced temperature difference. Reduced temperature difference is the ratio of the difference between water inlet temperature and ambient temperature to the intensity of solar radiation.

It can be seen from Figure 3(c) that system efficiency varies from a minimum of 36% to a maximum of 46%. Similarly, useful heat gain varies in between 1021W and 1490W, i.e. with a maximum of 1.5 kW. The experimental dataset generated through performance investigation of the ETSTC system (on ten different days) is given in Table 2. According to the Indian Standard IS 16368:2015 Clause no. 9.6.4, a curve (Figure 3c) is plotted with values of system efficiency ($\eta s_{ys)}$ on the y-axis and the values of reduced temperature (X) on the x-axis. After that, a trend line is drawn using the regression method

Figure 3a. Variation of solar insolation (Gt) and ambient temperature (Tad) with respect to the number of days

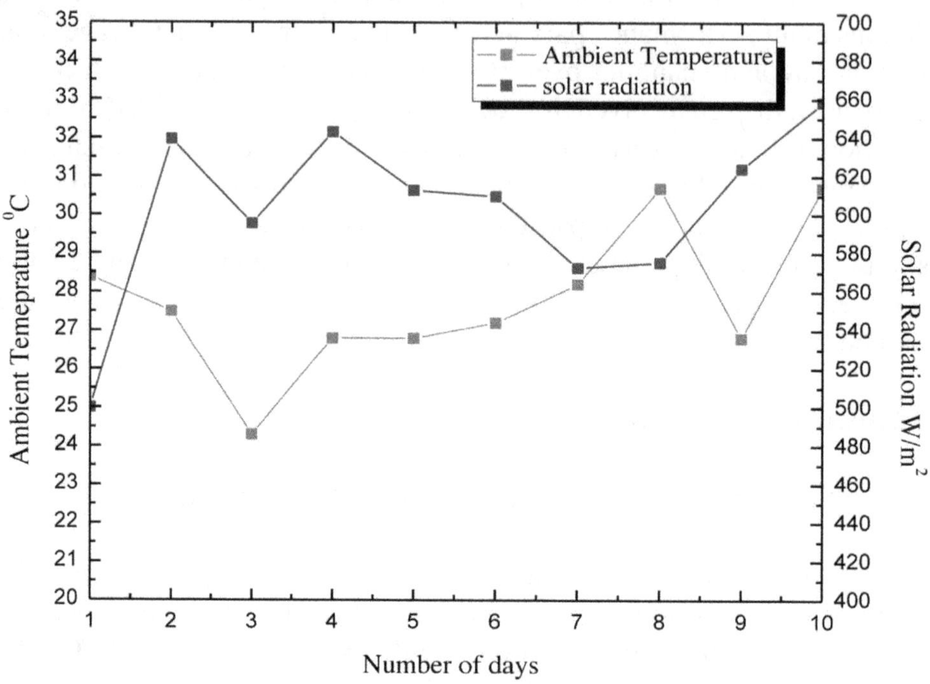

of curve fitting. The graph shows a fairly good correlation between the two parameters having R2 value close to 70%. Further, $\eta sy_{s\ d}$ecreases with the increase of X due to the negative slope of the line. This is attributed to the increase of thermal losses from the collector due to the increase of average ambient temperature of heat addition to the collector. The increase of the latter is the result of the increase of fluid temperature of the evacuated tubular collector. Therefore, it can be further noted that solar radiation, ambient temperature, and an end day temperature of the fluid (variables contained in X) have more influence on the performance of the evacuated tubular collector. This graph is one of the important performance characteristics of the ETSTC system as the system efficiency is more influenced by the three operating parameters, namely solar radiation, ambient temperature and an end day temperature of the working fluid. The maximum system efficiency of 46% is obtained at the fourth test run, i.e. on the 4th [a]verage solar insolation day (Table 1), which is obtained for solar insolation of 643 W/m2, the non-dimensional X value of 0.035, and ambient temperature 26.8 oC.

Figure 3b. Variation of useful heat gain and system efficiency (η) with respect to the number of days

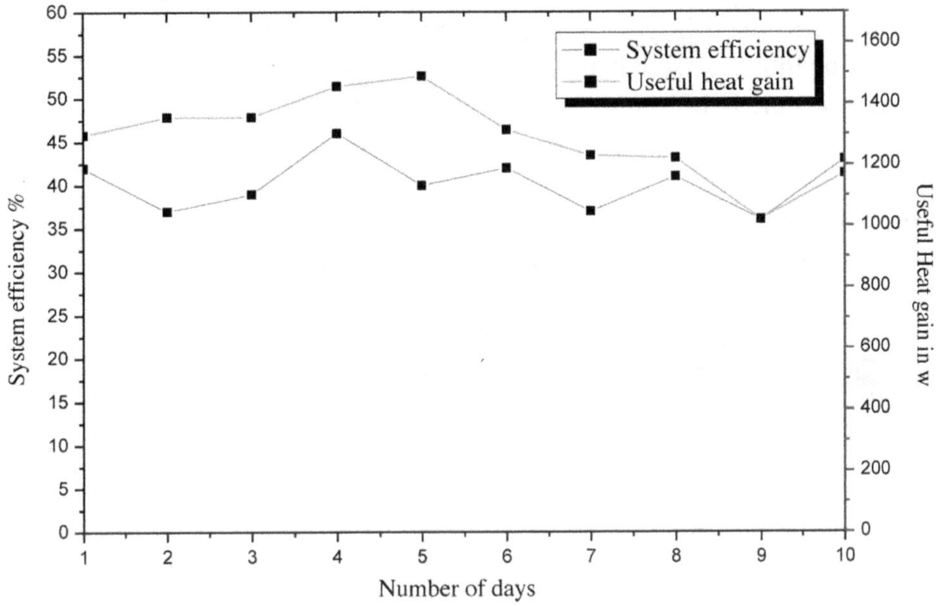

Figure 3c. Plot of ETSTC system efficiency with non-dimensional X

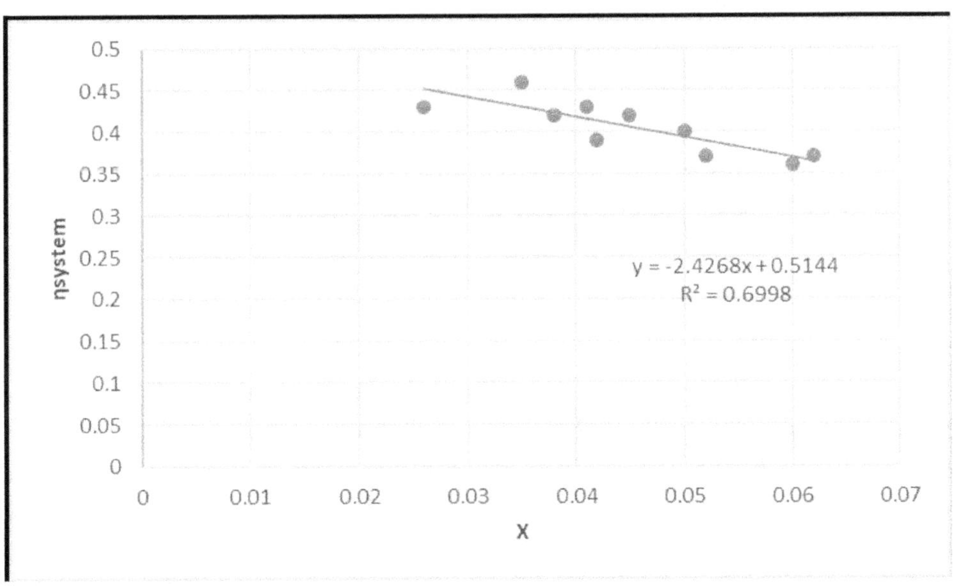

The performance of the ETSTC system can be improved by attaching parabolic mirror reflectors below the evacuated tubes. The proposed experimental setup is shown in Figure 4, which includes the ETSTC system in Figure 4(a) used in generating the experimental datasets, the parabolic mirror concentrators which are mounted underneath the tubes of the ETSTC system in Figure 4(b).

Figure 4. ETSTC system showing the set up diagram (a), parabolic mirror reflectors (b), and (c) schematic of how concentration takes place

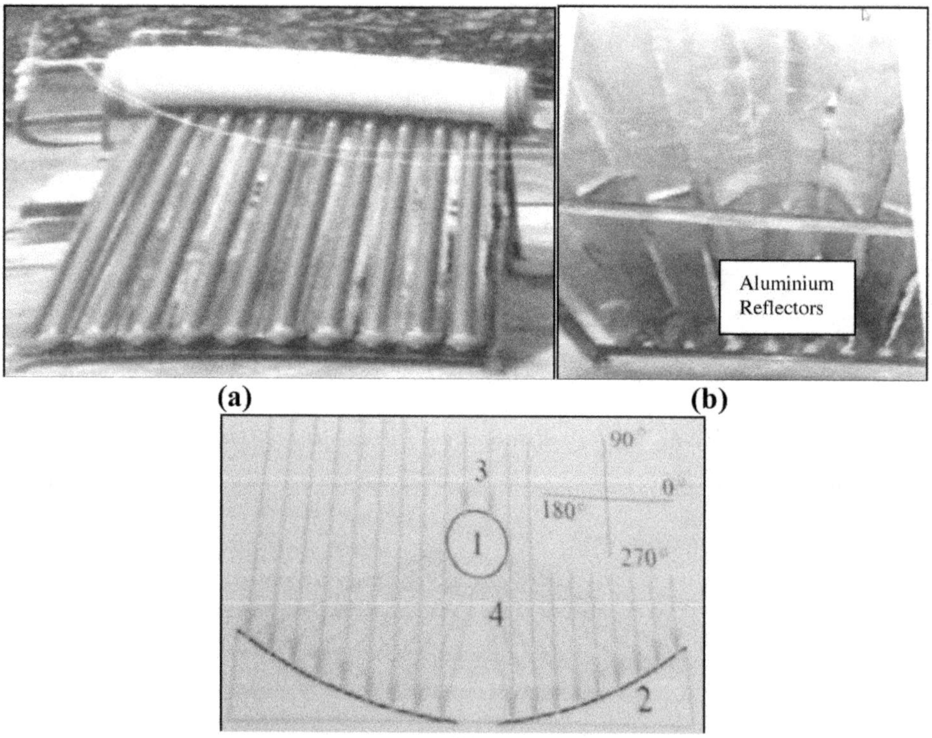

(a) (b)

(c)

In addition, how concentration of solar radiation takes place onto the receiver tube after getting incident on the parabolic mirror is shown in Figure 4(c). These reflectors are made from aluminium sheet bent into parabolic mirror shape. Aluminium is selected, for it has high reflectivity. Experimental datasets can be generated by performing the performance tests of such modified configuration of the ETSTC system. It is expected that the reflector below the tubes will be able to concentrate the otherwise lost solar radiation falling in the space between two consecutive evacuated tubes. Naturally when the diffused part of solar radiation falling on the empty space between any two consecutive tubes is utilized by concentrating the diffused rays onto the tubes with the help of reflectors, more increase of temperature of the working fluid will be possible. There are total ten number of ETSTC tubes used for the collector. The tubes are concentric two glass tubes with a vacuum sealed in the annulus. The outer diameter of the outer glass tube is 47 mm and the outer diameter of the inner glass tube is 37 mm. The aperture plane of the ETSTC is 1.5 m in length (i.e. length of the tubes) and 1 m in width. The tubes supply hot water after absorbing incident solar radiation to a cylindrical tank, which is of capacity 150 litres. In this tank the hot water is finally stored. The same experimental setup with reflector is used in generating the earlier experimental datasets as mentioned in Table 2. The present experimental datasets generated for analysing the ETSTC system performance with the parabolic mirror reflectors is given in Table 6. As can be seen from Table 6, the major output parameters are fluid outlet temperature and the ETSTC system efficiency, which are dependent on the input mass flow rate (operating parameter) and the ambient temperature, global solar radiation and wind speed (weather parameters). In order to monitor these variations, Figure 5

Table 6. Experimental results of Evacuated Tubular Solar Thermal Collector (ETSTC) with reflectors underneath the tubes

Test no.	Input Parameters				Output parameters	
	Mass flow rate (lpm)	Ambient temperature (T_{ad}) in °C	Global solar radiation (G_T) in W/m²	Wind speed W_s (m/s)	Outlet temperature (°C)	System Efficiency(η) in %
1	8.5	27	456	0.3	77.5	68.9
2	8.62	30	524	0.4	84	74.3
3	8.53	34	504	0.3	85	75.4
4	8.59	35	562	0.32	87.4	76
5	8.62	33	561	0.41	86	79.5
6	8.57	31	466	0.4	85	77
7	8.54	30	512	0.3	85.6	79.4
8	8.6	30	478	0.3	83	73
9	8.56	29	542	0.4	88.9	78.7

Figure 5. Variation of water outlet temperature and ETSTC system efficiency with mass flow rate

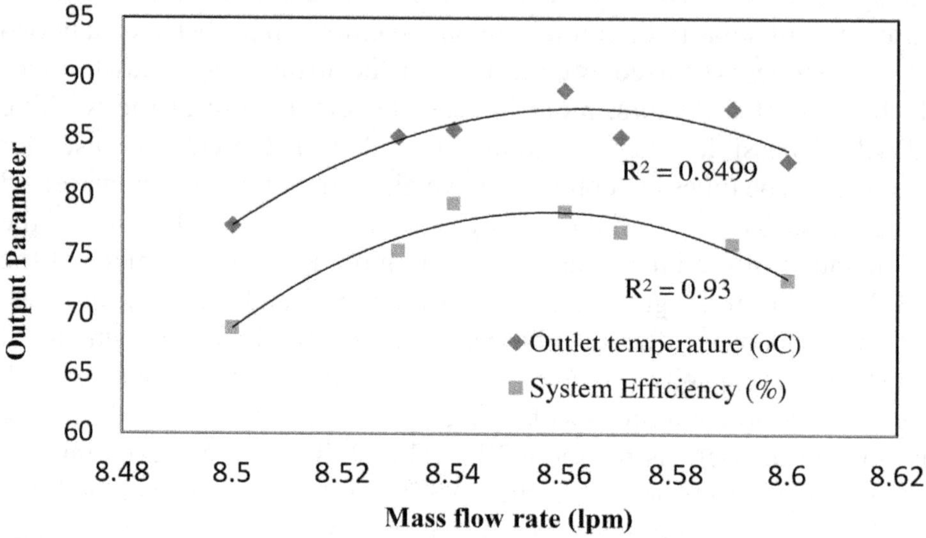

and Figure 6 show the variations of water outlet temperature and system

Figure 6. Variation of water outlet temperature and ETSTC system efficiency with global solar radiation

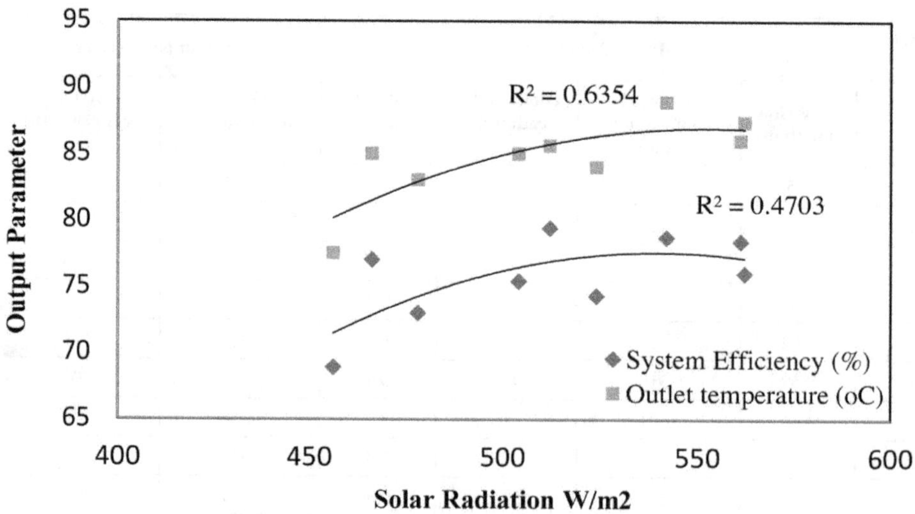

Figure 7. Standard efficiency plot of the ETSTC system with parabolic mirror reflectors

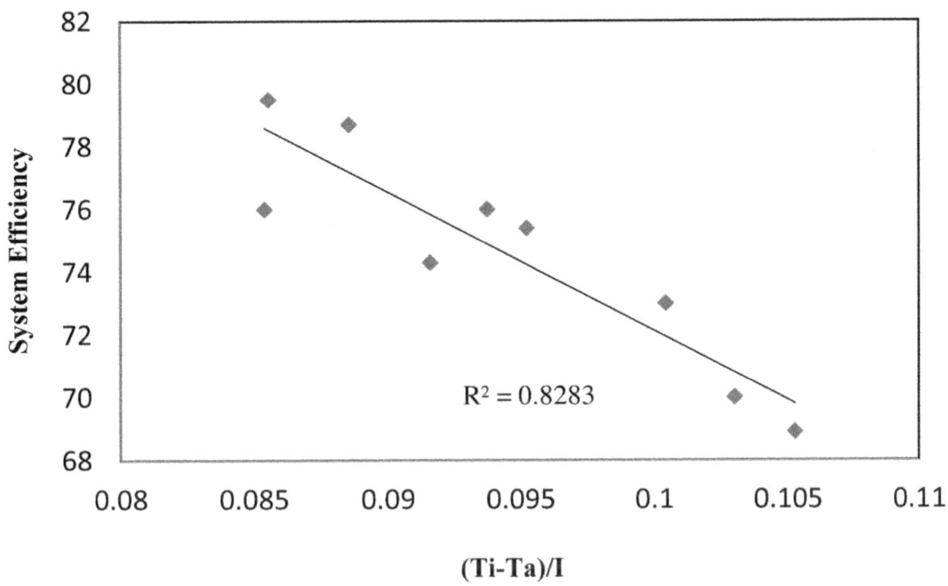

efficiency (i.e. thermal efficiency) of the ETSTC with respect to mass flow rate and solar radiation.

It can be seen from Figure 5 that with the increase of mass flow rate, the fluid outlet temperature and ETSTC system efficiency increase up to a certain extent and then both decrease with further increase of the former. Best fit second degree polynomial curves are fitted with degree of correlation (R-squared) values close to unity, from which these trends of variation are obtained. Hence there is an optimal mass flow rate at which outlet temperature and system efficiency are the maximum. A maximum outlet temperature of 88.9°C is obtained at an optimal mass flow rate of 8.56 lpm. And a maximum ETSTC system efficiency of 79.4% is obtained at an optimal mass flow rate of 8.54 lpm, which is very close to the optimal mass flow rate of maximum outlet temperature. Thus, increased mass flow rate does not necessarily increase the system performance beyond the optimal conditions. With respect to the weather parameter, i.e. solar radiation, Figure 6 shows that both the water outlet temperature and efficiency of the ETSTC system increases with the increase of solar radiation, but at a reduced rate at higher intensities. Therefore, there cannot be any optimal weather data of solar radiation for maximum performance of the ETSTC system; on the contrary it depends on

Figure 8. Standard efficiency plot of the ETSTC system with parabolic mirror reflectors

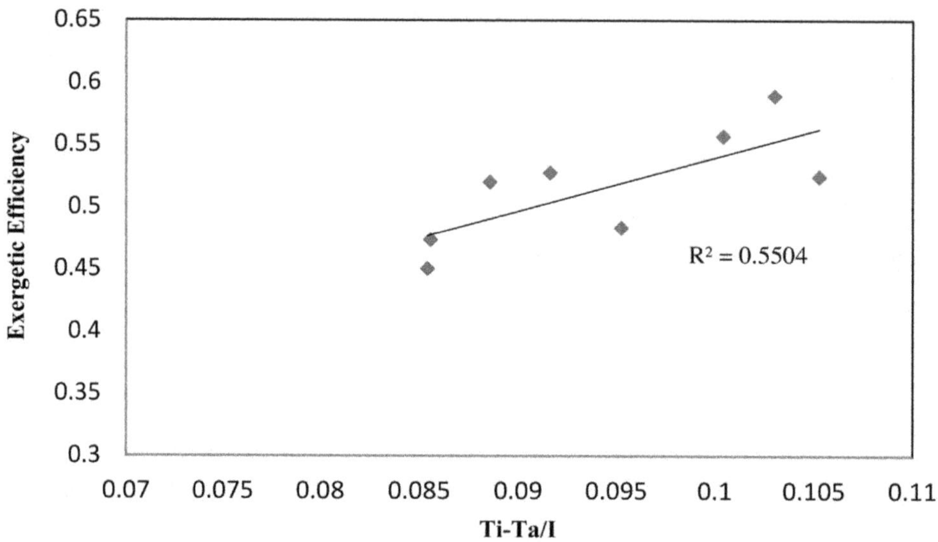

the operating mass flow rate to maximise the system performance. In this case, the performance is the maximum in terms of efficiency and water outlet temperature when the solar radiation is in the range 512-542 W/m^2.

Finally, the standard efficiency plot of the ETSTC system with parabolic mirror reflectors underneath is shown in Figure 7, which shows the variation of ETSTC system efficiency with reduced temperature difference. It follows the same trend as that of solar flat plate collector system efficiency. In comparison to Figure 3(c), it can be seen that the system efficiency is higher in case of ETSTC system with parabolic mirror reflector. A maximum system efficiency of 79.4% is obtained at a reduced temperature difference of 0.085. Figure 8 shows the variation of the exergetic efficiency of the ETSTC system having reflector with respect to the changes in reduced temperature difference. As also defined earlier, the reduced temperature difference is the ratio of the difference between water inlet temperature and the ambient temperature to the intensity of solar radiation. The ambient temperature is considered at an average value of 29°C, which is also the characteristic of the place where these datasets were generated. The expression for the calculation of exergetic efficiency is the same as given in the Chapter 2. It follows the same trend in the variation of exergetic efficiency as that for the solar flat plate collector. As the reduced temperature difference increases by the decrease of solar radiation or the increase of water inlet temperature to the collector, the exergy of water increases thereby leading to the increase of exergetic efficiency of the collector. Hence, with increase of reduced temperature difference, the exergetic efficiency increases whereas thermal efficiency increases. This is the result of the counteractive effect of reduced temperature difference on thermal and exergetic efficiencies. In Figure 8, a maximum exergetic efficiency of 0.589 is obtained at a reduced temperature difference of 0.103. In comparison to the value of reduced temperature difference at maximum thermal efficiency, the same at maximum exergetic efficiency is higher, which is counter-intuitive considering the trend in the variation of thermal efficiency and exergetic efficiency with respect to the reduced temperature difference.

Variation of Temperature

The ETSTC experimental trail runs at ten different days conducted, the temperature variation of water is considered starting day temperature and end day temperature. The temperature of water gradually increases with ten different days of experimental conduction. The temperature variation of water is minimum day 1 is 32.5°C starting day temperature and 48°C is end day

Figure 9a. Temperature variation of working fluid

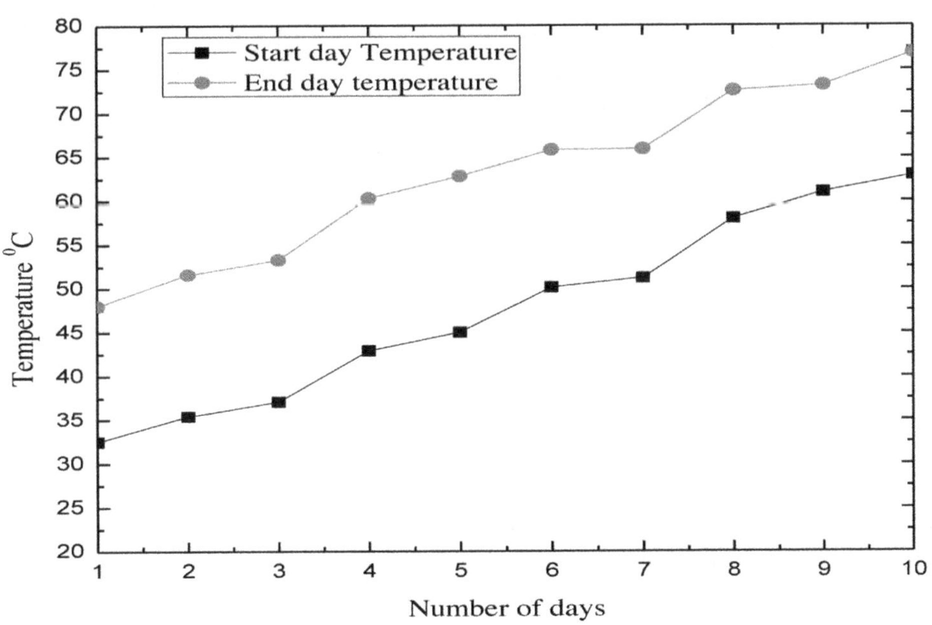

temperature. The maximum temperature of water is day 10 is 62.9°C starting day temperature and 76.9°C is end day temperature of the ECT system is shown in Figure 9(a). The temperature variation of the fluid is depends on the solar radiation. If solar radiation increases with day time at the same time the fluid temperature also increases. The temperature differences of ETSTC experimental investigation of ten different days are shown in Figure 9(b). The maximum temperature difference of working fluid is 17.8°C on 5th day of experimental run and minimum temperature difference of 12.2°C on 9th day of the experimental run. The variation of wind speed effect on the temperature difference of the working fluid.

Figure 9b. Temperature difference of working fluid

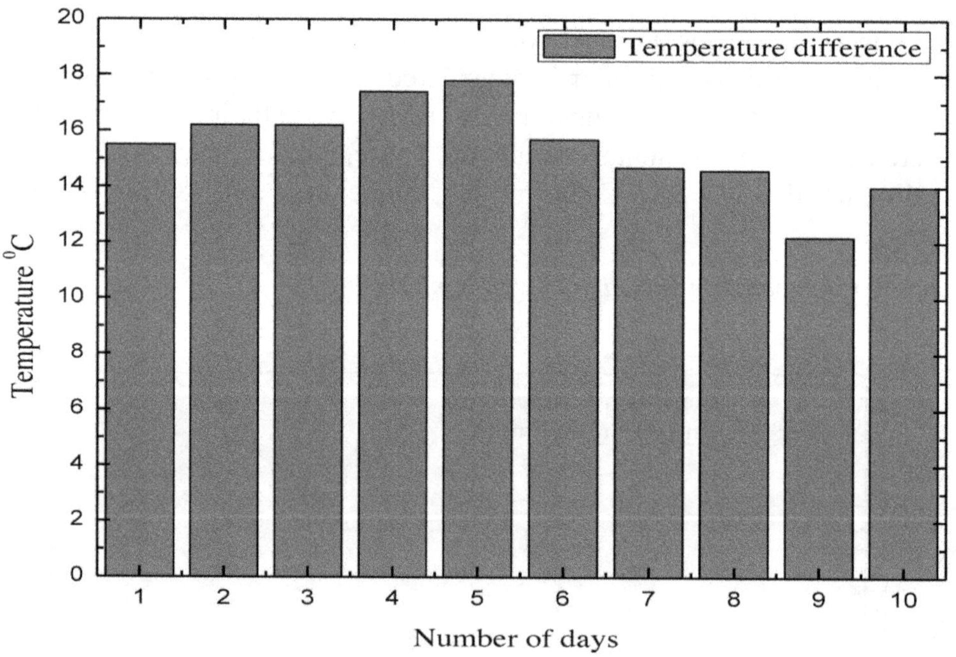

Variation of Wind Speed

In the present study the variation of wind speed is import parameter in the experimental investigation. The increases wind speed decreases the temperature of the working fluid. The concentration of solar radiation on the ETSTC system reduces due to high wind speed. In the present study considered the

Figure 10. Variation of wind speed

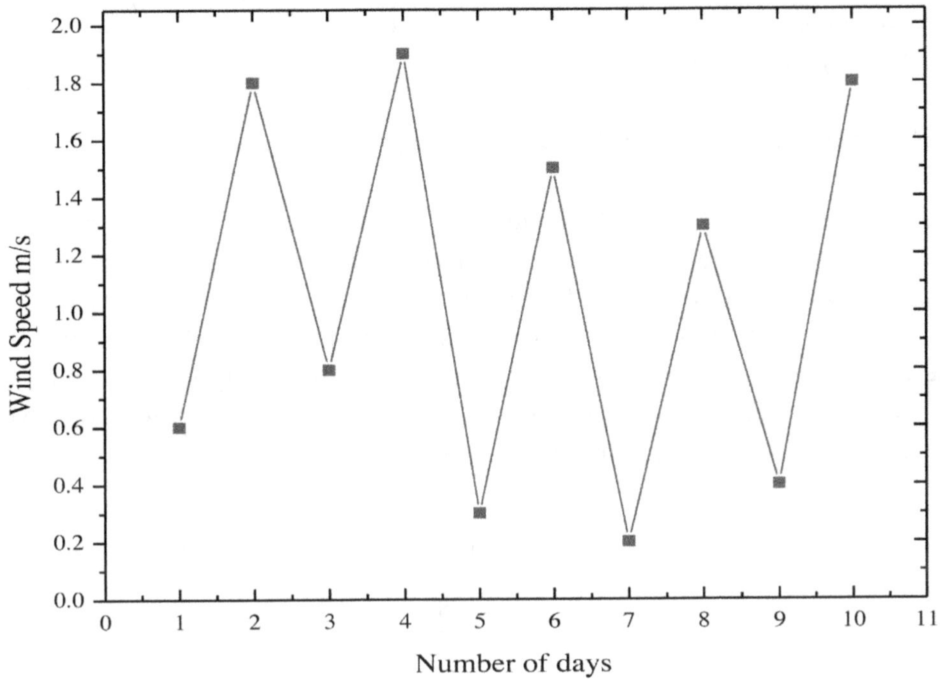

ten different days wind speed directly affect the performance of the system. The variation of wind speed is depicted in the Figure 10. The average low

Table 7. Relative index values of ETSTC system

Test No.	Relative Index (RI)	Rank
1	3.512	10
2	0.974	6
3	2.178	9
4	0.518	3
5	1.826	8
6	0.550	4
7	1.755	7
8	0.500	2
9	0.842	5
10	0.083	1

wind speed occurs on 5,7,9 days and average maximum wind speed on 2,4, and 10 days as shown in Figure 10. In the present study occurs the maximum wind speed is 1.9 m/s on 4th day experimental run and minimum wind speed is 0.2 m/s on 7th day experimental run.

Optimization of ETSTC Parameters

The optimization of ETSTC system is performed via M-SAW method considering five independent and two dependent parameters. Optimization results are already depicted in Table 4 and Table 5. Then, the ranking of the trail nos. is performed based on the Relative Index (RI) values (Table 7). The results show (Figure 11) that trail no.10 yields the highest rank with a relative index value of 0.083 and the trail no.1 yields least rank with relative index 3.512, which is the worst performance result.

Figure 11. Ranking of experimental results using a modified SAW method

10	0.083	1

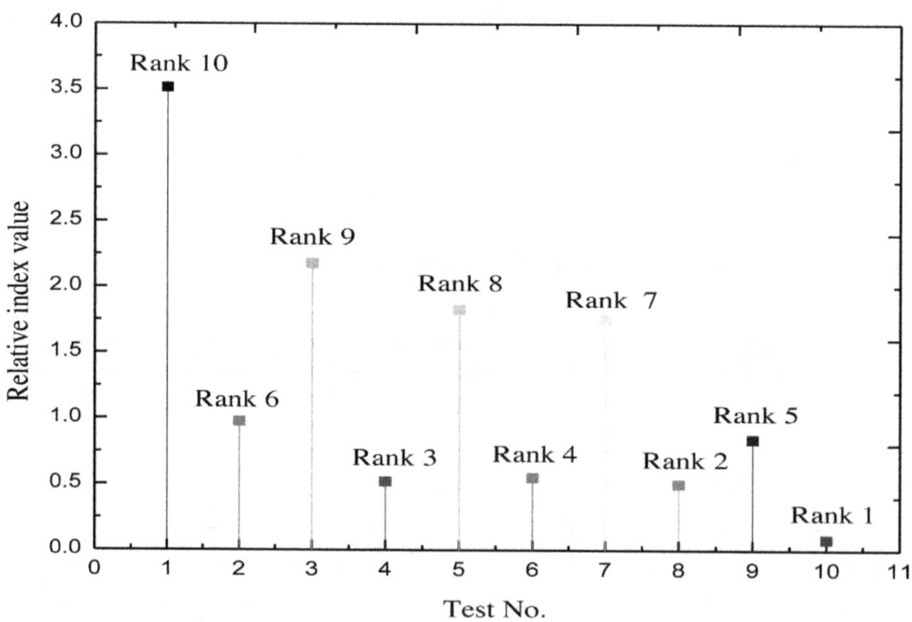

The optimal setting is obtained as T_{sid} = 62.9°C, T_{sfd}=76.9°C, T_{ad}=30.7°C, G_T = 658.4 W/m², and X=0.062, which gives optimal performance parameter i.e. system efficiency (η) as 43% and wind speed (W_s) as 1.8 m/s. The recommended optimal setting provides optimal performance like higher efficiency and higher wind speed which directly or indirectly increase the performance of ETSTC system.

Analysis of Variance Analysis

Further, Analysis of Variance (ANOVA) analysis with 95% of Confidence Interval is used to determine the significance and adequacy of the developed model. In the ANOVA analysis, the Fisher's statistical test (F-Ratio) and significance probability (P-value) followed by coefficient of determination (R^2) and adjusted-R^2 are determined to test the sufficiency and fitness of the proposed model. The F-value denotes the most influencing parameters on the performance characteristics while P-value shows the statistical significance of process variables of ETSTC system. During the analysis, process variables which have larger F-values and lower P-value ($P<0.05$) indicate their significance statistically (Kulekci, et al. 2011). The R^2 signifies the total variation in the response expected by the model. The purpose of adjusted-R^2 is to analyze the model adequacy and fitness. In this research stepwise elimination method is used to remove the insignificant parameters

Table 8. ANOVA for RI values of ETSTC system

Source	DF	Adj SS	Adj MS	F-Value	P-Value
Regression	4	9.5991	2.39977	78.56	0
Start day temperature (T_{sid})	1	2.0124	2.01241	65.88	0
global solar radiation (G_T)	1	1.0384	1.03842	33.99	0.002
non-dimensional reduced temperature (X)	1	0.4657	0.46568	15.24	0.011
wind speed W_s	1	0.1148	0.11481	3.76	0.11
Error	5	0.1527	0.03055		
Total	9	9.7518			
R^2=98.43%, R^2(adj)=97.18%, R^2 (Pred)= 88.46%					

to adjust the fitted quadratic model (Rao, & Kalyankar, 2014). ANOVA analysis results are given in Table.8.

The ANOVA for RI result shows (Table 8) that the parameter start day temperature (T_{sid}), global solar radiation (G_T), non-dimensional reduced temperature (X), wind speed (W_s) are the most significant factors influencing the assessment of RI values of ETSTC performance. However ambient temperature (T_{ad}) is considered as statistically insignificant, since it failed to test the significance. Subsequently, the P values for start day temperature (T_{sid}), global solar radiation (G_T), non-dimensional reduced temperature (X), wind speed (W_s) are found to be less than 0.05, which indicate that the model and associated terms are statistically significant (Konig, et al., 1985). The other parameters except parameters highlighted in Table 4.8 are considered to be statistically insignificant. The value of R^2 is found out to be 0.9843 for RI values, which implies that 95% of experimental data are well-suited. Higher R^2 coefficient indicates satisfactory model to the experimental data. Finally, the adequacy and fitness of the model is calculated by adjusted-R^2 values. The high value of adjusted-R^2 (0.8846 for RI values) supports a high correlation between the experimental and the predicted values (Kumar, et al. 2014).

Confirmatory Test

At last, confirmatory tests are conducted to verify or validate the optimal setting obtained via the proposed method. The optimal setting i.e. T_{sid}=62.9°C, T_{sfd}=76.9°C, T_{ad}=30.7°C, G_T = 658.4 W/m², X=0.062, and W_s = 1.8 m/s are used for confirmatory experiments and the corresponding results shown in Table 9. The results show that confirmatory tests results are comparable and acceptable with experimental results for the optimal conditions.

Table 9. Confirmatory tests for SFPC thermal system

Input Conditions	Output Parameters	Experimental results	Confirmatory results
Test No-10: T_{sid} = 62.9°C, T_{sfd}=76.9°C, T_{ad}=30.7°C, G_T = 658.4 W/m², X=0.062, and W_s = 1.8 m/s.	T_{sfd} in °C	76.9	75.3
	η in %	43.0	42.5

SUMMARY

Modeling and optimization of ETSTC using modified SAW method is discussed. Experiments (10 nos.) have been carried out on ETSTC by varying five independent parameters [start day temperature (T_{sid}), end day Temperature (T_{sfd}), average air temperature (T_{ad}), global solar radiation (G_T), non-dimensional parameter (X)]. Investigation results show that test no.10 is the optimal experimental trail run and corresponding collector efficiency is obtained as 43%. Subsequently, T_{sid}, T_{sfd}, G_T, T_{ad} and W_s for trail no. 10 is 62.9 °C, 76.9 °C, 658.4 W/m^2, 30.7 °C and 1.8 m/s, respectively. Further, parametric analysis results show that parameters such as solar radiation, ambient temperature, and end day temperature of the fluid have more significant effect on the ETSTC than the others. Additionally, ANOVA analysis is carried out to determine the significance and adequacy of the developed model. The results show that response parameters start day temperature (T_{sid}), global solar radiation (G_T), non-dimensional reduced temperature (X), wind speed (W_s) are the major parameters for ETSTC system and also model fitted data are well fitted with experimental results within the 95% confidence level. At last, confirmatory tests show comparable and acceptable results with experimental values. Finally, as the MCDM technique used give optimized values of the considered operating parameters and efficiency of the collector, and as these are obtained with good accuracy and with statistical significance, the obtained optimized values can be practically implemented for the present solar collector. Therefore, it is concluded that performance of ETSTC under different input conditions is found to be feasible and proposed method provides an acceptable and satisfactory results.

REFERENCES

Abrams, W., Ghoneim, E., Shew, R., Lamaskin, T., Al-bloushi, K., & Hussein, S. (2018). Delineation of groundwater potential (GWP) in the northern United Arab Emirates and Oman using geospatial technologies in conjunction with Simple Additive Weight (SAW), Analytical Hierarchy Process (AHP), and Probabilistic Frequency Ratio (PFR) techniques. *Journal of Arid Environments*, 1–20. doi:10.1016/j.jaridenv.2018.05.005

Afshari, A., & Mojahed, M., & Yusuff, R.M. (2010). Simple Additive Weighting approach to Personnel Selection problem. *International Journal of Innovation, Management and Technology*, *1*, 511–515.

Ashrae. (2010). *Methods of Testing to Determine the Thermal Performance of Solar Collectors*. Academic Press.

Bornatico, R., Pfeiffer, M., Witzig, A., & Guzzella, L. (2012). Optimal sizing of a solar thermal building installation using particle swarm optimization. *Energy*, *41*(1), 31–37. doi:10.1016/j.energy.2011.05.026

Hassan, Z., & Al, G. (2017). A Fuzzy AHP and GIS-based Approach to Prioritize Utility-Scale Solar PV Sites in Saudi Arabia. *IEEE International Conference on Systems, Man, and Cybernetics (SMC)*.

Hassan, Z., Al, G., & Anjali, A. (2017). Solar PV power plant site selection using a GIS-AHP based approach with application in Saudi Arabia. *Applied Energy*, *206*, 1225–1240. doi:10.1016/j.apenergy.2017.10.024

Hayek, M. (2009). Investigation of evacuated-tube solar collector's performance using computational fluid dynamics. *Proceedings of*, *ACTEA09*, 240–244.

Hill, J. E., & Streed, E. R. (1976). A method of testing for rating solar collectors based on thermal performance. *Solar Energy*, *18*(5), 421–429. doi:10.1016/0038-092X(76)90008-6

Jagannath R., Biplab D., & Jagadish. (2019). Study on Effect of Barriers in Green Supply Chain Management Using Modified SAW Technique: A Case Study. *Optimizing Current Strategies and Applications in Industrial Engineering*, 202-219.

Kadek, A. W., Riyanarto, S., & Nurul, F. A. I. (2016). Optimization Solar Farm Site Selection Using Multi Criteria Decision Making Fuzzy AHP and PROMETHEE: Case Study in Bali, International Conference on Information. *Communication Technology and System (ICTS)*, 237-243.

Kalogirou, S. A. (2004). Optimization of solar systems using artificial neural networks and genetic algorithms. *Applied Energy*, *77*(4), 383–405. doi:10.1016/S0306-2619(03)00153-3

Kalogirou, S. A. (2004). Solar thermal collectors and applications. *Progress in Energy and Combustion Science*, *30*(3), 231–295. doi:10.1016/j.pecs.2004.02.001

Kek, S. P., Chin, N. L., Tan, S. W., Yusof, Y. A., & Chua, L. S. (2018). Comparison of DNA extraction methods for entomological origin identification of honey using simple additive weighting method. *International Journal of Food Science & Technology*, *53*(11), 2490–2499. doi:10.1111/ijfs.13840

Kim, Y., & Seo, T. (2007). Thermal performance comparisons of the glass evacuated tube solar collectors with shapes of the absorber tube. *Renewable Energy*, *32*(5), 772–795. doi:10.1016/j.renene.2006.03.016

Kim, Y., & Seo, T. (2007). Thermal performances comparisons of the glass evacuated tube solar collectors with shapes of absorber tube. *Renewable Energy*, *32*(5), 772–795. doi:10.1016/j.renene.2006.03.016

Konig W., Wulf, C. & Willerscheid H. (1985) . Machining of fibre reinforced plastics. *CIRP Annals Manufacturing Technology*, *34*, 537-548.

Kulekci, M. K., Esme, O., Er, O., & Kazancoglu, Y. (2011). Modeling and prediction of weld shear strength in friction stir spot welding using design of experiments and neural network. *Materialwissenschaft und Werkstofftechnik*, *42*(11), 990–995. doi:10.1002/mawe.201100781

Kumar, R., Kumar, K., & Bhowmik, S. (2014). Optimization of mechanical properties of epoxy based wood dust reinforced green composite using taguchi method. *International Conference on Advances in Manufacturing and Materials Engineering*, *5*, 688-696. 10.1016/j.mspro.2014.07.316

Mehrabi, D., Mohammadzaheri, M., Firoozfar, A., & Emadi, M. (2017). A fuzzy virtual temperature sensor for an irradiative enclosure. *Journal of Mechanical Science and Technology*, *31*(10), 4989–4994. doi:10.100712206-017-0947-x

Mevlu, U. (2013). GIS-based solar farms site selection using analytic hierarchy process (AHP) in Karapinar region, Konya/Turkey. *Renewable & Sustainable Energy Reviews*, *28*, 11–17. doi:10.1016/j.rser.2013.07.042

Mohanty, S., Rout, A., Patra, P. K., & Sahoo, S. S. (2017). ANFIS based solar radiation data forecasting for energy and economic Study of solar water heaters in eastern India. *International Journal of Control Theory and Applications*, *10*, 179–190.

Morrison, G. L., Budihardjo, I., & Behnia, M. (2004). Water-in-glass evacuated tube solar water heaters. *Solar Energy*, *76*(1-3), 135–140. doi:10.1016/j.solener.2003.07.024

Morrison, G. L., Budihardjo, I., & Behnia, M. (2005). Measurement and simulation of flow rate in water- in-glass evacuated tube solar water heater. *Solar Energy*, *78*(2), 257–267. doi:10.1016/j.solener.2004.09.005

Proctor, D. (1984). A generalized method for testing all glasses of solar collectors, II: Evaluation of collector thermal constants. *Solar Energy*, *32*(3), 387–394. doi:10.1016/0038-092X(84)90283-4

Rao, R.V., & Kalyankar, V.D. (2014). Optimization of modern machining processes using advanced optimization techniques—a review. *International Journal of Advanced Manufacturing Technology*, 1159-1188.

Suman, D., & Jagannath, R. (in press). Investigation of thermal performance of SAC using Fuzzy logic based expert system. *Journal of Mechanical Science and Technology*.

Walker, A., Mahjouri, F. & Stiteler, R. (2004). *Evacuated-Tube Heat-Pipe Solar Collectors Applied to the Recirculation Loop in a Federal Building*. NREL/CP-710-36149.

Zambolin, E., & Del, C. D. (2010). Experimental analysis of the thermal performance of flat plate and evacuated tube solar collectors in stationary standard and daily conditions. *Solar Energy*, *84*(8), 1382–1396. doi:10.1016/j.solener.2010.04.020

Zhang, X. R., & Yamaguchi, H. (2008). An experimental study on evacuated tube solar collector using supercritical CO2. *Applied Thermal Engineering*, *28*(10), 1225–1233. doi:10.1016/j.applthermaleng.2007.07.013

Chapter 5
Modeling and Optimization of Parabolic Trough Collector

ABSTRACT

Parabolic trough collector (PTC) is a concentrating collector widely used for steam cooking, water heating, and also steam power generation and desalination work. The performance of PTC is strongly depends on its process parameters and is a MCDM problem. Implementation of integrated method, that is, entropy with graph theory and matrix approach (E-GTMA) for modelling and optimization of PTC parameters to improve higher outlet temperature (To), higher heat gain (h), and higher thermal efficiency (ηth), is discussed in this chapter. Investigation results indicate the effectiveness of this technique for multi-objective optimization and determined optimal setting as Test no.10 for PTC. Additionally, parametric and ANOVA analysis is carried out to determine the significance and adequacy of the developed model. Last, validation of the proposed model and verification results is done via confirmatory tests, and tests results show comparable and acceptable w.r.t. experimental results.

INTRODUCTION

Parabolic trough collector is a concentrating collector of line focusing type widely used for steam cooking, water heating, and also steam power generation and desalination work (Lupfert, et al. 2011). Concentrated solar energy is a promising source of energy that is currently attracting many targeted efforts

DOI: 10.4018/978-1-7998-3523-3.ch005

to enhance its exploitation. In particular PTC for low enthalpy processes is an emerging technology. This technology converts the solar beam radiation into useful thermal energy at the receiver unit. The PTC consists of an absorber tube, a concentric transparent cover (in order to prevent convective heat loss by maintaining thermal insulation over the absorber tube surface), and a parabolic reflector. The absorber, also called receiver tube is fixed permanently at the focus of the parabolic reflector. The generation of thermal energy for some industrial processes requires temperatures in the range 80°–400°C. And the temperature of the PTC absorber tube can be as high as 350°–400°C. The parabolic trough collector is placed on a rigid structure and the Sun tracking mechanism is placed on the rigid structure to track the Sun rays by the parabolic trough (or reflector). Sun tracking is an important feature of commercial PTC systems. This is done to ensure that solar radiation falls normally, i.e. at zero incidence angle, onto the aperture plane of the parabolic trough.

The first parabolic trough collector was built in the year 1936, which produced a power of 0.37 kW through a steam engine (Spencer, 1989). The parabolic trough is made from a polished sheet bent into a parabolic shape. The receiver tube is made up of copper or stainless steel having good absorptivity and less emissivity. The tube is sometimes enclosed within an outer transparent cylindrical cover to prevent thermal loss due to convection and re-radiation. It can also be coated with selective coating that has a high absorptance for solar radiation, but low emittance for thermal radiation loss. When the parabolic trough is pointed towards the Sun, solar radiation incident on the parabolic trough is reflected onto the receiver tube located at the focal plane of the trough. The working fluid (here it is water) when passes through the tube gets heated up due to conduction and convective heat transfer (i.e. conjugate heat transfer), and then the hot water is stored in the insulated hot water storage tank. Hot water is stored in the storage tank by means of thermal stratification of water, which means that the hot water temperature is lower in the bottom layers and higher in the top layers, i.e. to say that hot water temperature increases from bottom to top layers. The intermixing of hot water in the storage tank destroys the thermal stratification and after mixing uniform temperature distribution can be obtained in the hot water storage tank. The PTC system is called a line-focusing system if the collector tube is assumed to be replaced with a line at the focal plane. The parabolic trough can be oriented along East-West direction so that it can track the Sun during its traverse from East to West in the sky due South direction for any location in the Northern hemisphere, like India. The tracking can be a single axis or multi-axis tracking type (Roy, et al., 2016). Multi-axis tracking is required for point focusing systems. However, point focusing requires a complicated electronic circuitry for tracking the Sun in all directions. Obviously the cost also increases with the complex construction of the tracking system.

On the other hand, a programmed motorized unit can provide single-axis tracking. After the design of PTC and investigation of PTC performance is also essential and it depends on the various independent and dependent parameters. Improper selection of these parameters results in an error in the results which directly and indirectly the performance of PTC. Therefore, optimization of PTC parameters is essential to get higher outlet temperature (T_o), higher heat useful gain (*H*) and higher thermal efficiency (η_{th}). Further, optimization of PTC is considered to be an MCDM problem because PTC involves any number of independent parameters like alternatives and the number of dependent parameters like criteria.

Although many investigators have worked on optimization of other thermal systems using techniques like ANN (Kalogirou, 2004), PSO (Bornatico, et al., 2012), ANFIS (Mohanty, et al., 2017), Fuzzy Set theory (Mehrabi, et al., 2017), GRA (Ebrahimi, & Rahmani, 2018), Fuzzy-GRA (Rao, & Padmanabhan, 2007) etc., it is observed that studies are limited to single objective optimization and few are multi-response optimization techniques. However, extraction of precise priority weights for each of the system parameters, which influence the performance, is ignored. Also, consideration of both beneficial and non-beneficial parameter in the estimation of the system performance is missing. Thus, MCDM based optimization technique is essential for modeling and optimization of PTC. In the present chapter, a hybrid integrated MCDM method is proposed which consists of entropy with GTMA method. Further, no literature on modeling and optimization of PTC using entropy-GTMA method is available. The present integrated method i.e. entropy-GTMA provides many advantages like flexibility to solve multi-criteria problem, considers multiple criteria during optimization with objective data, easy to maintain consistency, easy priority weights calculation, less computational time required, very simple and stable technique, minimum mathematical calculation involved, which may be quite useful and helpful to the decision makers who may not have a strong background in mathematics.

Experimental Details

The experiments on the PTC system were conducted for a typical sunny day from 9.30 am to 14.20 pm at a given mass flow rate and the reading was taken at an interval of 10 minutes (as given in Table 1) A photograph of the laboratory-scale experimental test rig, installed at the Regional Test Center, National Institute of Technology Silchar, India is shown in Figure 1. A suitable schematic diagram of the system with all its subsystems is shown in Figure 2. As can be seen from these figures, an experimental test rig of the PTC system consists of the following subsystems- a parabolic trough (of standard length 6 feet and perimeter 4 feet) usually made of stainless steel sheet of thickness 2 mm; copper receiver tube at the focal plane having external diameter 1 inch and internal diameter 0.98 inch; a puff insulated hot water storage tank (HWST) made from aluminum with fiber reinforced plastic (FRP) lining for rust prevention. The thickness of puff insulation is 2 cm. Water enters the receiver tube at one end and comes out from the other end as a hot fluid, which is then stored in the storage tank. Hot water is stored in the storage tank after thermal stratification, which means hot water is stored in layers inside the tank having the hot most layers on the top i.e. having a temperature gradient in the upward direction. The capacity of the storage tank is 28 liters.

The PTC system used in this work has a concentration ratio of 45.83. Concentration ratio is defined as the ratio of the aperture area of the parabolic trough to the external surface area of the receiver tube. For Sun tracking of the trough, an East-West single axis tracking mode is used, which is controlled by a micro-controller. The micro-controller gives signal to a stepper motor, which rotates incrementally by 0.004° per step. The tracker can turn from 40° due East to 40° due South. Since the earth rotates by 15° per hour (according to hour angle) in the device (Figure 1), the motor can be programmed such that it rotates the trough exactly by 15° in an hour, thereby having Sun's normal incidence on the aperture plane of the parabolic reflector or trough (Roy, et al., 2016). For taking experimental readings various instrumentations are used for measuring the ambient and operating parameters as shown in Figure 2, for e.g. an electronic flowmeter (range 0.5-25 lpm, accuracy ± 1%) for measuring flow rate in the absorber tube; differential thermostat for set point temperature; PT 100 type RTD sensors for reading stagnation temperature (range 0 to 150°C, accuracy ± 0.5°C) at inlet and outlet of the collector; solar pyranometer for measuring intensity of solar radiation (0-2000W/m^2, accuracy ± 1W/m^2) circulating pump (fractional horsepower, i.e. 0.5 hp) and water supply unit. The flowmeter passes controlled flow rate as per the standard operating conditions. The solar radiation incident on the aperture plane is measured using solar pyranometer, which is mounted in the aperture plane.

The experimental datasets were generated after augmenting the performance of the receiver tube, which was coated with a selective coating to improve absorptance of solar radiation. Before that both the receiver tube and the reflecting parabolic surface were cleaned with acid pickling to remove any accumulated dust and then washed with cold water. The coating was prepared by mixing titanium white powder with spirit to form a paste, which was applied externally on the tube surface with the help of a brush so as to drape the tube surface uniformly with the coating. Titanium paste exhibited good adhesiveness when applied on the tube surface. With coating the tube diameter was measured with a micrometer and its difference with the initial tube diameter was calculated to find the thickness of the coating, which was maintained 5mm. The necessary shape calibration of the coating was also ensured to have uniform selective coating on the tube surface. After the coating was set, experimental readings were taken. The time gap between each reading was set at 10 minutes. The machine was started (both circulation and the tracker) at least half-an hour prior to documenting the first reading. Table1 consists of different trial numbers, which correspond to the data point's equivalent to the time periods at the considered interval of time.

Integrated E-GTMA method

An integrated entropy-GTMA is proposed for modeling and optimization of PTC. Here, Entropy method (Rao, 2006) is employed for extraction of precise priority weights while GTMA for optimization of PTC parameters with considering the interrelationship aspects of PTC criterias and alternatives. The schematic flow chart of E-GTMA is shown in Figure 3. The details steps of E-GTMA method (Rao, 2006; Kalogirou, 2020; Kulekci et al., 2011; Rao, & Kalyankar, 2014; Konig, et al., 1985; and Kumar, et al., 2014) are explained as follows:

Step-1: Identification of process attributes

The proposed method starts with the identification of process criteria's, sub-criteria's, and alternatives required for evaluation of the performance of PTC. These criteria's/alternatives consist of both qualitative and quantitative parameter as well as beneficial and non-beneficial criteria's which influence the PTC performance.

Step-2: Formulation of decision matrix and normalization

Figure 1. Experimental setup of parabolic trough collector (PTC) at National Institute of Technology Silchar, India

In this step, based on the identified process attributes, the formulation of the decision matrix is carried out using the Equation (1). The decision matrix contains performance values of the number of criteria's (each column elements) and a corresponding number of alternatives (each row elements) (Rao, 2006; Kalogirou,). After that, normalization of the decision matrix is carried out to convert the different measurement performance values to comparable sequence values using Equations (2-3). In the normalization process, higher values for beneficial while lower values for non-beneficial criteria are desirable.

Figure 2. Schematic diagram of a parabolic trough collector (PTC)

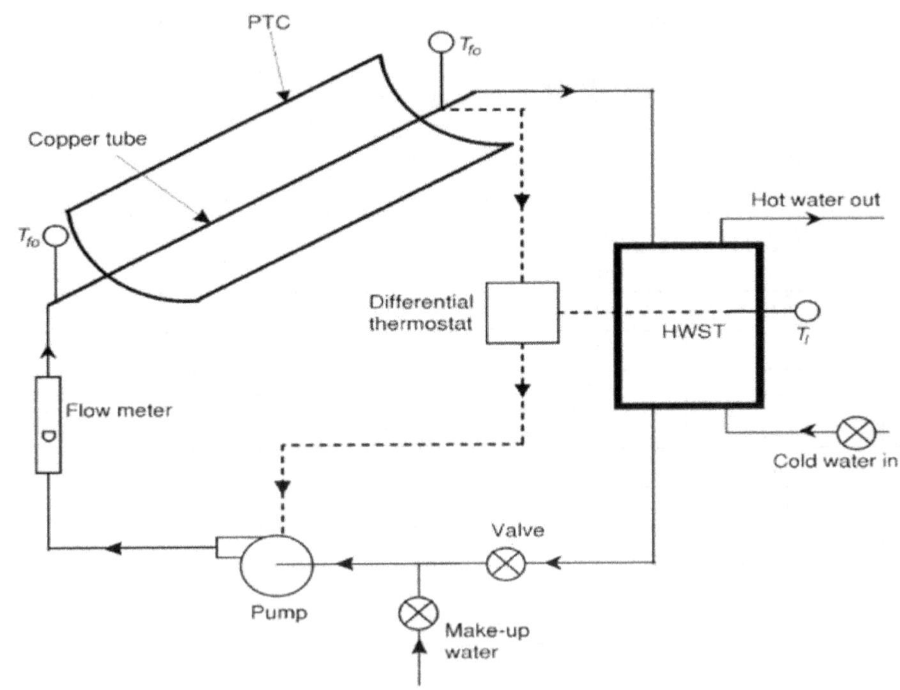

$$D_{ij} = \begin{bmatrix} & C_1 & C_2 & \dots & C_n \\ A_1 & a_{11} & a_{12} & \dots & a_{1n} \\ A_2 & a_{21} & a_{22} & \dots & a_{2n} \\ \dots & \dots & \dots & \dots & \dots \\ A_m & a_{m1} & a_{m2} & \dots & a_{mn} \end{bmatrix} \tag{1}$$

where, C_1, C_2, ... C_n represent the number of criteria's, A_1, A_2, A_3 ...... A_m represent the number of alternatives, a_{11}, a_{12},...... a_{mn} denote performance values of each criteria corresponding to each alternative, n is the number of criterias and m denotes the number of alternatives in the process.

$$[N_{ij}] = \frac{D_{ij}}{D_{bigger}} \tag{2}$$

Table 1. Experimental results of PTC with selective coating of the receiver tube

Trial No.	Input parameters		Output parameters		
	Inlet temperature (T_i) in °C	**Solar Radiation (I) in W/m²**	**Outlet temperature (T_o) in °C**	**Useful heat gain (H) in Watt**	**Thermal efficiency(η_{th}) in %**
1	26.3	538.26	42.4	376.04	36.2
2	27.5	564.46	43.3	391.06	35.7
3	28.5	589.55	44.1	415.29	34.3
4	29.3	613.70	45.1	426.73	33.7
5	30.1	631.87	45.8	451.22	32.6
6	31.1	652.11	46.9	468.31	30.7
7	31.7	672.05	47.9	461.72	30.6
8	32.5	688.33	48.6	495.92	29.8
9	33.4	697.86	49.4	503.82	29.5
10	34.1	535.60	46.3	312.02	29.4
11	34.5	558.96	47.7	317.79	30.9
12	36.0	578.70	48.8	325.55	29.3
13	37.4	595.44	51.0	340.97	30.7
14	38.7	611.82	52.2	350.71	29.4
15	40.0	616.77	53.3	355.03	29.1
16	41.2	637.42	53.9	363.64	26.9
17	42.3	652.52	53.2	371.02	22.3
18	43.3	659.06	54.1	378.31	21.5
19	44.3	662.77	55.2	384.41	21.0
20	45.5	677.69	56.0	385.36	20.3
21	46.2	688.01	56.8	393.39	19.3
22	47.1	691.38	57.6	397.11	21.0
23	47.9	679.88	58.4	389.82	20.5
24	48.6	665.77	59.0	383.6	19.3
25	49.3	649.03	61.2	364.93	22.9
26	49.9	638.75	61.7	351.56	24.7
27	50.1	623.29	62.2	335.99	26.4
28	50.7	609.40	63.1	325.64	24.9
29	51.5	589.97	64.1	315.67	25.8
30	52.3	571.99	64.6	310.62	25.4
Constant Parameters: Mass flow rate (m) = 0.0031 kg/s; Absorber Area (A) = 0.319 m²					

Figure 3. Schematic flow chart of E-GTMA method for PTC

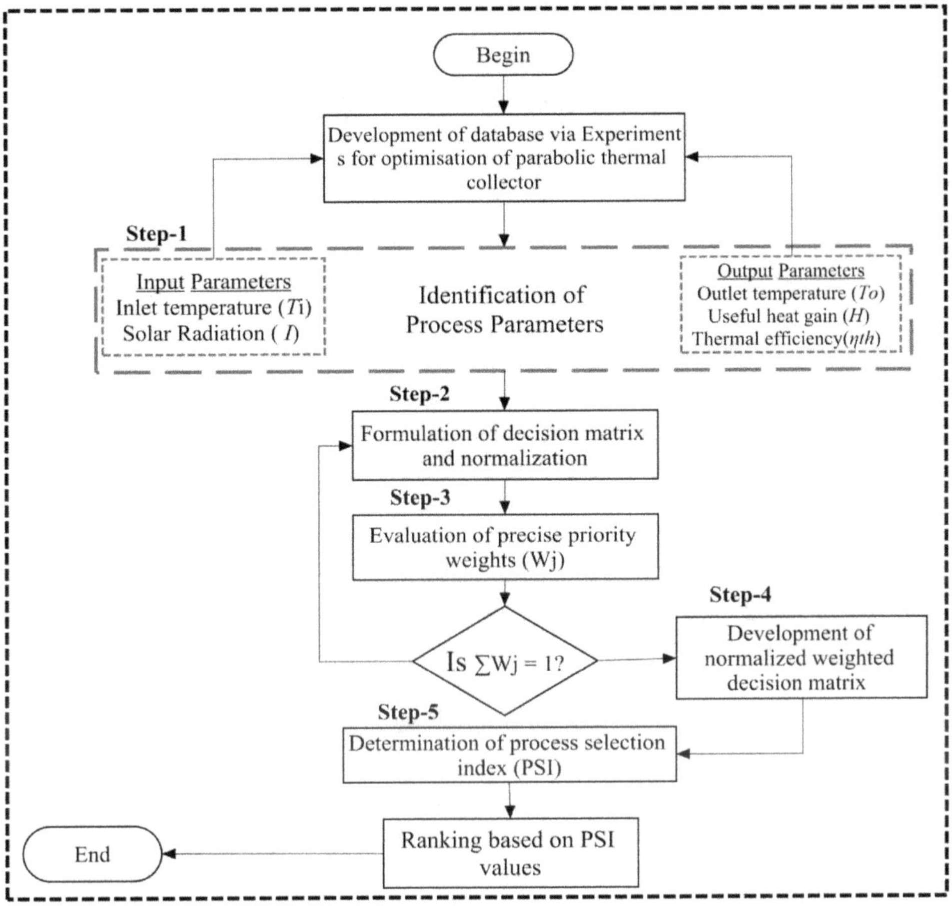

$$[N_{ij}^*] = \frac{D_{smull}}{D_{ij}} \tag{3}$$

where, $[N_{ij}]$ and $[N_{ij}^*]$ are the normalized decision matrix for beneficial and non-beneficial criteria; D_{bigger} is the beneficial element of a column and $D_{smaller}$ is the non-beneficial element of a column (Rao, 2006; Rao, & Kalyankar, 2014).

Step-3: Evaluation of precise priority weights

After the normalization, precise priority weights for each of the criterias are determined using Equations (4-6). First, the degree of divergence (d_j) for each of the criterias is calculated, which measures the distance of each criteria with respect to the comparable sequence. After that precise priority weights are determined using Equation (6).

$$R_j = -k\sum_{i=1}^{m}\left([N_{ij}]and[N_{ij}^*]\right)\ln\left([N_{ij}]or[N_{ij}^*]\right) \quad j = 1,2,\ldots,n. \tag{4}$$

$$d_j = \left|1 - \Re_j\right|. \tag{5}$$

$$\zeta_j = \frac{d_j}{\sum_{j=1}^{n} d_j}. \tag{6}$$

where, R_j is the entropy matrix; k is a constant term and its value is determined by [k=1/ln m, $0 \le d_j \le 1$], d_j denotes the degree of divergence for each of the criterias, ζ_j are the precise priority weights for each of the criterias.

Step 4: Development of normalized weighted decision matrix

In this step, the priority weights obtained by using Equation (6) are multiplied with the normalized performance values of each criterion in the decision matrix (Rao, 2006; Kalogirou, 2020). The normalized weighted decision matrix is calculated using the following expression.

$$W_{ij} = \zeta_j \times \left([N_{ij}]and[N_{ij}^*]\right) \tag{7}$$

where, W_{ij} represents the weighted normalized decision matrix, ζ_j are the precise priority weights for each of the criteria's and $[N_{ij}]$ and $[N_{ij}^*]$ are the normalized decision matrix for beneficial and non-beneficial criterias.

Step 5: Determination of process selection index (PSI) and ranking

Process Selection Index (PSI) of each alternative is calculated by just adding the normalized weighted performance values for each criterion. This converts the multi-attribute problem to a single attribute problem with considering the effect of beneficial and non-beneficial criteria's of the process into account (Rao, 2006; Kalogirou, 2020). Thereafter, the ranking of PSI values is done in ascending order; and the ranks of each individual alternative are obtained. The optimal alternative for the process is the one which has maximum PSI.

Modeling of PTC System

Modeling of PTC is presented using entropy coupled with GTMA method (Rao, 2006). Here, first, identification of process attributes i.e. process/input parameters and corresponding performance/output parameters of the PTC is done (Kumar, et al., 2014). Work considered two input parameters i.e. inlet temperature (T_i) in °C and solar radiation (I) in W/m^2 while three output parameters i.e. outlet temperature (T_o) in °C, useful heat gain (H) in Watt and thermal efficiency (η_{th}) in % (Table 1). Second, development of decision matrix is done using Equation (1). In the present work, the decision matrix is developed via experimentation in PTC. Total 30 trials are performed for different input conditions and experimental results as tabulated in Table 1. Then, normalization of the data is carried out to convert the original sequence data into comparable sequence data using Equations (2-3) and the results of the normalization is shown in Table 2. Third, determination of priority weights of output parameters i.e. T_o, h and η_{th} are done using entropy method (Kalogirou, 2020; Kulekci et al., 2011). In this, the first reference range is calculated using Equation (4), which measures the target value of entire PTC decision matrix data. After that, the determination of the distance of each PTC parameter from the reference value i.e. degree of divergence (d) and precise priority weights (ζ) are evaluated using Equations (5-6). The results show that parameter W (0.3988) yields higher weights signifies higher impact on PTC performance compare to the T_o, η_{th}.

After the weights determination, the formulation of a normalized weighted decision matrix is carried out using Equation (6), in which obtained weights using Equation (6) for each of the PTC parameters are multiplied with normalized performance values of corresponding PTC parameters. Similarly, PSI values for each of the PTC parameters i.e. T_o, h and η_{th} are done by adding normalized weighted performance values for each criterion. This converts the multi-attribute problem into a single attribute problem with considering the effect of beneficial and non-beneficial criteria of the process (Kulekci et al., 2011; Rao, & Kalyankar, 2014). At last, the ranking of PSI values is done in ascending order; the ranks of each individual alternative are obtained.

Table 2. Normalized output parameters and priority weights of PTC

Trial. No	Outlet temperature (°C)	Useful heat gain (W)	Thermal Efficiency (%)
1	0.0266	0.0329	0.0445
2	0.0272	0.0342	0.0439
3	0.0277	0.0363	0.0421
4	0.0283	0.0373	0.0414
5	0.0287	0.0394	0.0400
6	0.0294	0.0409	0.0377
7	0.0301	0.0403	0.0376
8	0.0305	0.0433	0.0366
9	0.0310	0.0440	0.0362
10	0.0290	0.0273	0.0361
11	0.0299	0.0278	0.0380
12	0.0306	0.0284	0.0360
13	0.0320	0.0298	0.0377
14	0.0327	0.0306	0.0361
15	0.0334	0.0310	0.0357
16	0.0338	0.0318	0.0330
17	0.0334	0.0324	0.0274
18	0.0339	0.0331	0.0264
19	0.0346	0.0336	0.0258
20	0.0351	0.0337	0.0249
21	0.0356	0.0344	0.0237
22	0.0361	0.0347	0.0258
23	0.0366	0.0341	0.0252
24	0.0370	0.0335	0.0237
25	0.0384	0.0319	0.0281
26	0.0387	0.0307	0.0303
27	0.0390	0.0294	0.0324
28	0.0396	0.0285	0.0306
29	0.0402	0.0276	0.0317
30	0.0405	0.0271	0.0312
Priority weights of output parameters for PTC			
Weights (ζ)	0.3144	0.3988	0.2868

RESULTS AND DISCUSSION

Based on the detailed modeling of the PTC system using the proposed MCDM technique, system optimization has been done for improving the performance of the same. But before that a detailed parametric analysis is done using an experimental approach to elucidate how the different input parameters affect the output parameters.

Parametric Analysis

The performance of the present PTC system with receiver tube selective coating is evaluated in terms of the variation of its thermal efficiency and the useful heat gain over a time period equivalent of its heating period. Dynamic performance estimation (i.e. variation of performance with respect to time) of solar collector is required as the ambient parameters like solar radiation and ambient temperature are not constant over a given time period. Thus, the variation of ambient temperature and the intensity of solar radiation falling on the collector, also called solar irradiation, are monitored during the heating period as they impact the performance of the collector. The experimental investigation of PTC system was carried out at a technical university named National Institute of Technology Silchar, which is located in a city called Silchar in North-East of India.

Figure 4 shows the variation of useful heat gain and thermal efficiency over the heating period. The variation of solar irradiation is also plotted over the considered period of time. Almost continuous set of data points have been recorded over the day. The maximum value of heat gain is 503.82 W and minimum value of heat gain is 310.62 W with constant parameters of mass flow rate (m) = 0.0031 kg/s; absorber area (A) = 0.319 m^2. The maximum thermal efficiency is 36% and the minimum thermal efficiency is 25%. It can be seen that variations of solar radiation and useful heat gain by the collector follow the same trend as there is a direct influence of solar radiation on the heat gain. As can be seen from Figure 5, there is a sudden drop in solar radiation and also useful heat gain in the period from 10:50 a.m. to 11:00 a.m., which might have been caused due to the sudden appearance of dwelling clouds. The thermal efficiency decreases with the increase of solar radiation as the effect of solar radiation on the thermal efficiency is counteractive. When solar radiation rises sharply (like the present case), thermal efficiency decreases due to increased thermal loss from the collector. As can be seen from Figure

Figure 4. Plot of solar irradiation, useful heat gains and thermal efficiency of PTC over time during a day

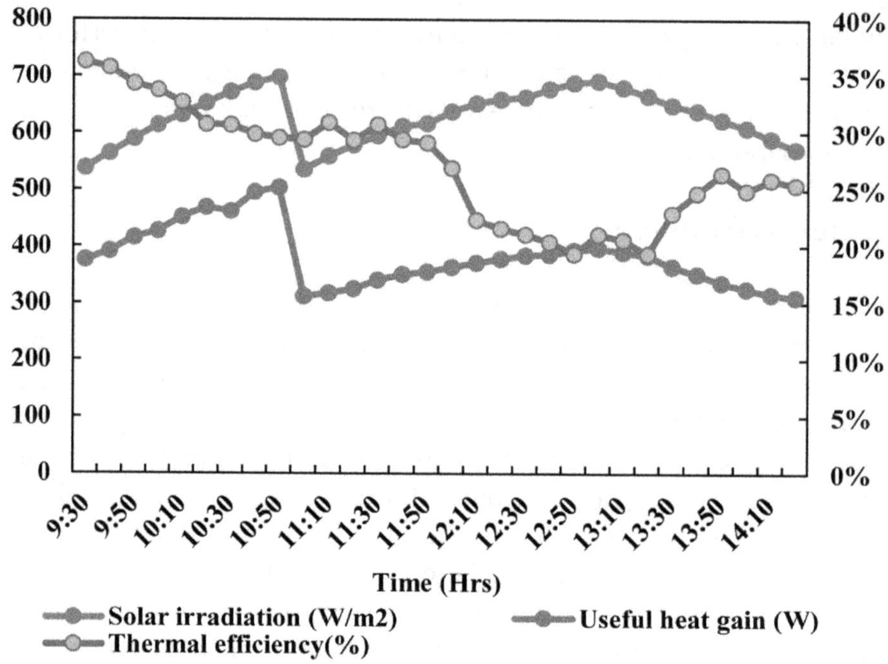

4, there are a few spikes in the readings of thermal efficiency around noon time and late afternoon. The reason for this is the result of the decrease in solar radiation during these periods.

Figure 5 shows the variation of water inlet temperature with respect to time during the heating period. The variation of thermal efficiency is also included in the same plot. It shows that the thermal efficiency decreases with the increase of inlet temperature to the collector. It is the result of increased thermal losses from the collector as the mean temperature of heat addition increases with the increase of water inlet temperature. These qualitative variations of performance parameters with respect to ambient and operating parameters follow the physics of the solar thermal collectors (Konig, et al., 1985). The variations of outlet and inlet temperatures of PTC system are shown in Figure 6. The outlet temperature of system is higher than the inlet temperature of the working fluid. This temperature variation of the fluid is depending on solar radiation. If the solar radiation increases the temperature

Figure 5. Variation of inlet temperature and thermal efficiency of a PTC during a day

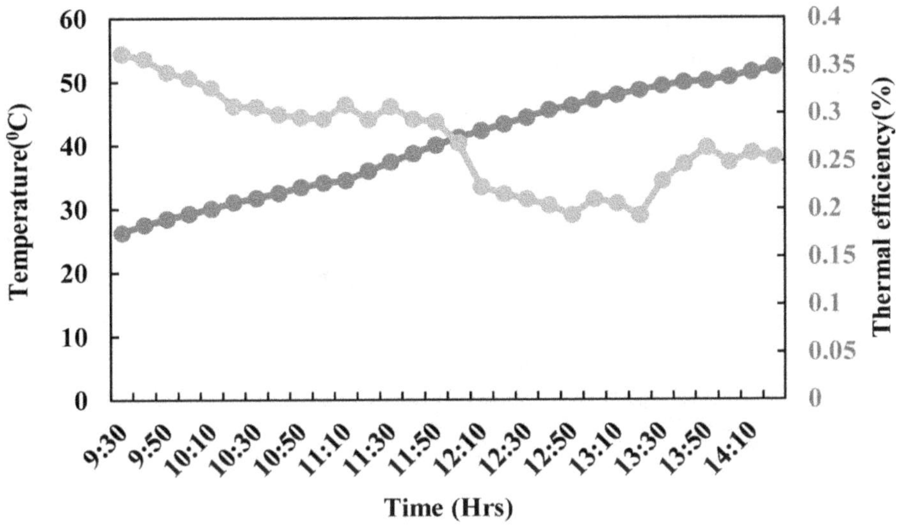

increases. The maximum temperature of outlet is 64.6°C at the end of the day and the minimum temperature is 42.4 °C at starting of the day.

According to the Indian Standard IS 16648: Part 5, 2017 (Roy, et al., 2016), a curve (Figure 7) is plotted with values of thermal efficiency on the y-axis and the reduced temperature difference (Ti-Ta/I) on the x-axis. This graph shows the composite effect of water inlet temperature and solar radiation for a given day ambient temperature (all these accounted for in the reduced temperature expression) on the thermal performance of the collector. It clearly shows that efficiency decreases with the increase of the reduced temperature caused by the thermal loss from the collector. Therefore, the graph in Figure 7 is one of the important performance characteristics of the PTC system, which can also be considered as the ultimate performance characteristic curve of the PTC system. A best-fit trend line is obtained using a regression method of least squares of curve fitting, which shows a degree of correlation (R^2 value) close to 61%.

As can be seen from Figure 7, the maximum thermal efficiency of about 0.36 is obtained for a reduced temperature difference of 0.0001. Figure 7 shows the variation of the exergetic efficiency of the PTC system with respect to the changes in reduced temperature difference. As already defined earlier, the reduced temperature difference is the ratio of the difference between

Figure 6. Variation of Inlet and Outlet temperature of fluid

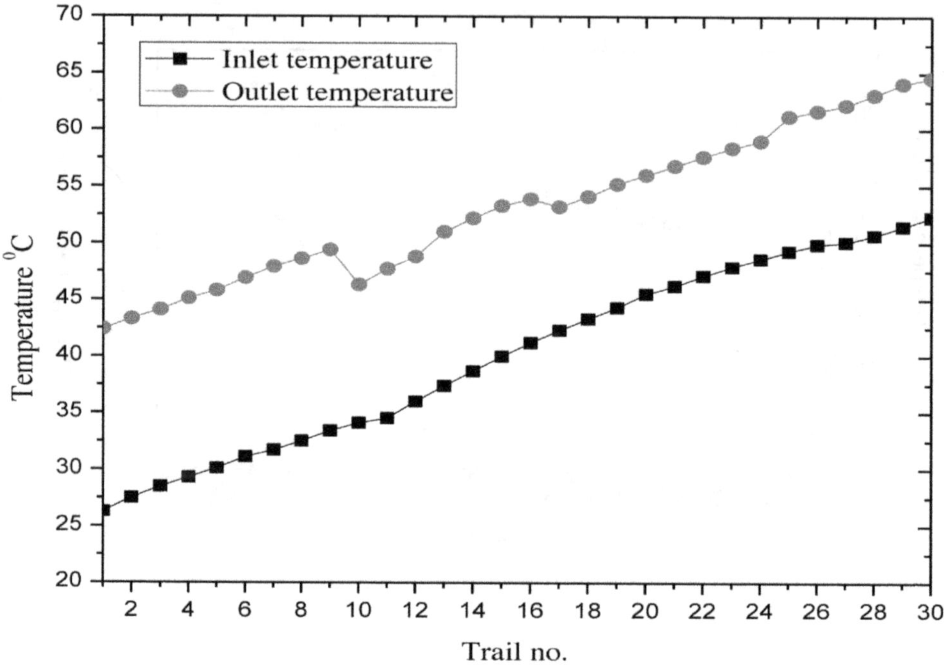

Figure 7. The plot of PTC system thermal efficiency with reduced temperature

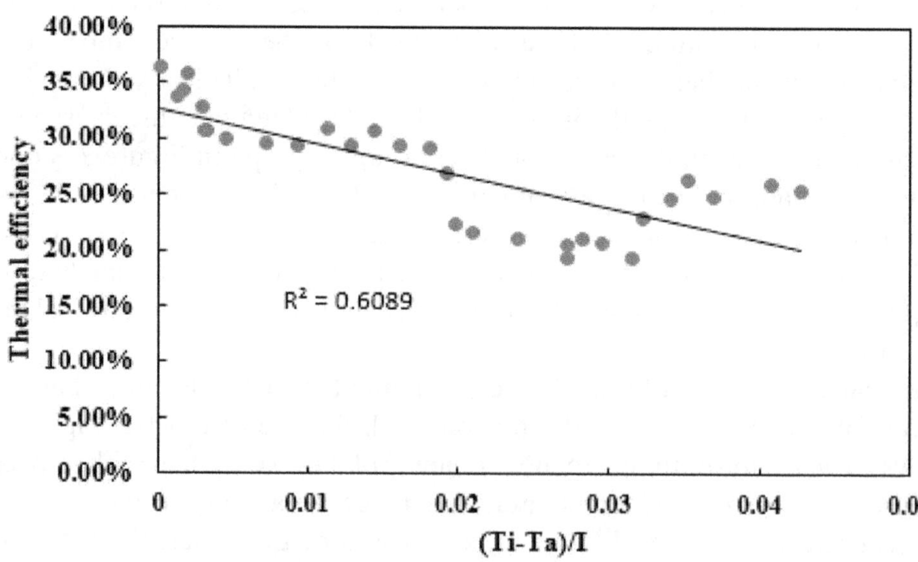

water inlet temperature and the ambient temperature to the intensity of solar radiation. The ambient temperature is considered at an average value of 29°C, which is also the characteristic of the place where these datasets were generated. The expression for the calculation of exergetic efficiency is the same as given in the Chapter 2. It follows the same trend in the variation of exergetic efficiency as that for the solar flat plate collector. As the reduced temperature difference increases by the decrease of solar radiation or the increase of water inlet temperature to the collector, the exergy of water increases thereby leading to the increase of exergetic efficiency of the collector. Hence, with increase of reduced temperature difference, the exergetic efficiency increases whereas thermal efficiency increases due to the counteractive effect of reduced temperature difference on these efficiencies. In Figure 8, a maximum exergetic efficiency of 0.094 is obtained at a reduced temperature difference of 0.062. Again the reduced temperature difference at maximum exergetic efficiency is higher than that at maximum thermal efficiency. This is counter-intuitive considering the trend in the variation of thermal efficiency and exergetic efficiency with respect to the reduced temperature difference (Roy, et al., 2016).

Figure 8. The plot of PTC system exergetic efficiency with reduced temperature

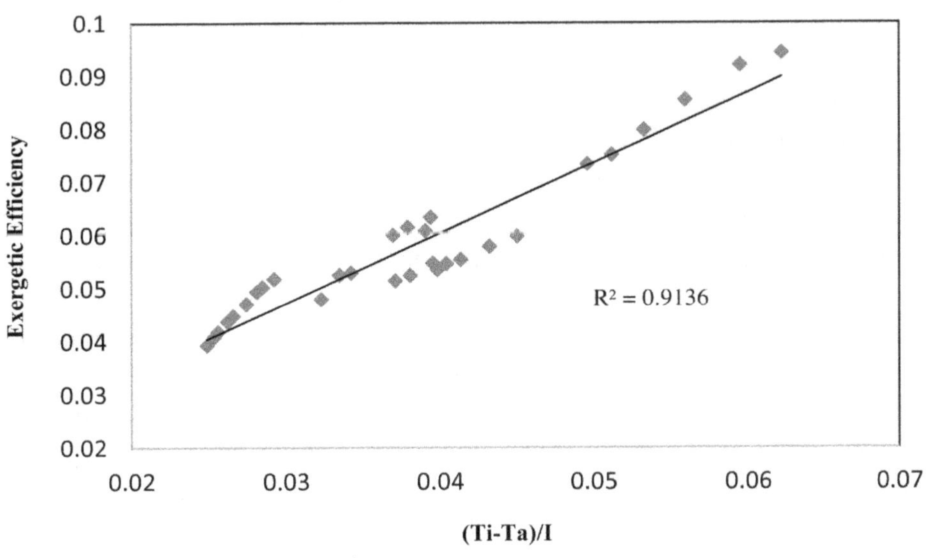

The performance of the PTC system can be improved by heat transfer augmentation of the working fluid in the receiver tube, which can be made possible by various means for example by using circular enclosure around the receiver tube to reduce the convective and re-radiation losses to the atmosphere, by using selective coating on the surface of the receiver tube to improve heat absorption (as discussed earlier), and by using twisted tape inserts in the receiver tube thereby increasing heat conduction. The schematic of the twisted tape inside a parabolic trough receiver tube is shown in Figure 9. In this plot, 'y' is the pitch distance (y), and AA is a cross-section of the tube. The right side view of this figure shows the inner and outer diameter of the tube. Both the conductive and the convective heat transfer coefficients are expected to increase due to the insertion of the twisted tape (Roy, et al., 2016). The augmentation of heat transfer rates by the insertion of twisted tape inside the tube is a passive means such that the flow attains a spiral motion as it flows in the axial direction, thereby modifying the Reynolds, and the Nusselt numbers. With the change in the pitch distance, the relative effects of Nusselt number and Reynolds number will be attenuated. However, once twisted insert is put inside, the pumping power may increase and the cost of operation might become high during the operation. Experimental dataset can also be generated like previous through performance investigation of the PTC system using twisted tape insert inside the receiver tube. Such dataset is given in Table 3 from which it can be seen that for the considered input parameters, i.e. mass flow rate, solar radiation and inlet temperature to the collector, the thermal efficiency of the PTC system has an average value of 0.49.

However, the variation of thermal efficiency of the PTC system having twisted tape inserts with respect to solar radiation and the same with respect to mass flow rate are shown in Figure 10 and Figure 11 It can be seen from these figures that there are optimal values of solar radiation and mass flow rate at which thermal efficiency is the maximum. The maximum thermal efficiency of 0.496 is obtained at an optimal mass flow rate of 5.7 kg/s and at an optimal solar radiation of 555 W/m^2. This value of the solar radiation is characteristic of almost average yearly solar radiation of the place where these experimental datasets are generated. In these two plots the second degree curve fitted through the data points has a degree of correlation (R-squared) value of unity (Roy, et al., 2016). The variation of thermal efficiency of the PTC system having the twisted tape inserts with respect to reduced temperature difference is shown in Figure 12. As also defined earlier, the reduced temperature difference is a characteristic of the collector behaviour, which is defined as the ratio of the difference between water inlet temperature and ambient temperature to the intensity of solar radiation. As can be seen, the trend in the variation of thermal efficiency is exactly similar to that of other solar thermal collectors as seen in the previous chapters. Here the curve fitting (straight line) is obtained with a degree of correlation (R-squared value) of 0.829. In comparison to the thermal efficiency variation of the PTC system (Figure 6) without twisted tape insert with respect to reduced temperature difference, it can be seen that the performance with twisted tape insert is much higher than without twisted tape insert. With twisted tape insert, a maximum thermal efficiency of 0.496 is obtained at a reduced temperature difference of 0.0149, whereas without twisted tape insert (Figure 6), a maximum thermal efficiency of 0.362 is obtained at a reduced temperature difference of 0.0001. In the present case, the values of reduced temperature difference have increased compared to previous case, which is due to the result of twisted tape insert (Roy, et al., 2016).

Table 3. Experimental results of PTC with twisted tape insert

Trial No.	Input parameters			Output Parameter
	Mass flow rate in kg/s	Solar Radiation (I) in W/m²	Inlet temperature in °C	Thermal efficiency (η_{th}) in %
1	5.7	555	39.3	0.496
2	6.8	500	42.7	0.488
3	7.6	590	46	0.49

Optimization of PTC Parameters

The PSI values for each of the trial nos. or alternatives are employed for finding the optimum operating conditions. The optimal alternative is the one which has maximum PSI and the results are depicted in Table 4. The ranking of PSI values is done in ascending manner and the corresponding result is shown in Figure 13.

The results show that trial no.10 yields higher PSI value with rank 1. This indicates that trial no. 10 is the optimal operating condition compared to the other trial nos. The optimal setting is obtained as T_i= 34.1°C, I=535.60 W/m² at a constant m = 0.0031 kg/s; and A = 0.319 m², which gives optimal performance parameter i.e. T_o= 46.3 °C, H= 312.03 W, and η_{th} = 29.4%,

Figure 9. Schematic of twisted tape inside a parabolic trough receiver tube

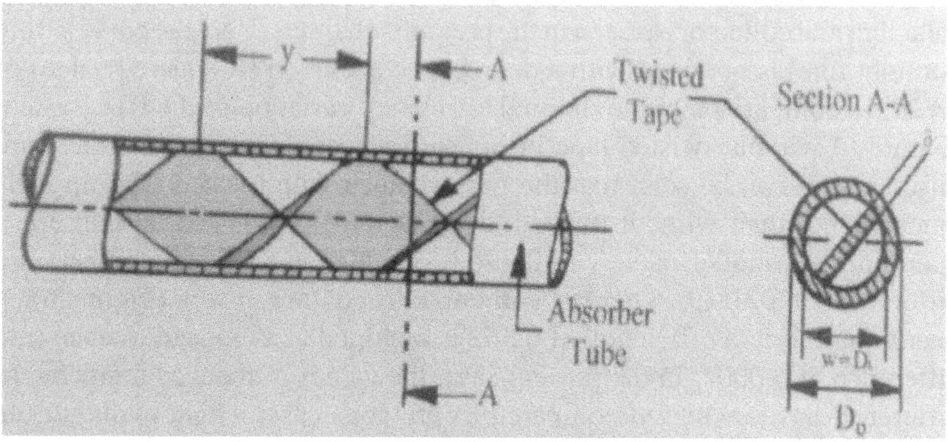

which directly or indirectly increases thermal efficiency (η_{th}) as well as the performance of PTC.

Figure 10. Variation of PTC thermal efficiency with global solar radiation

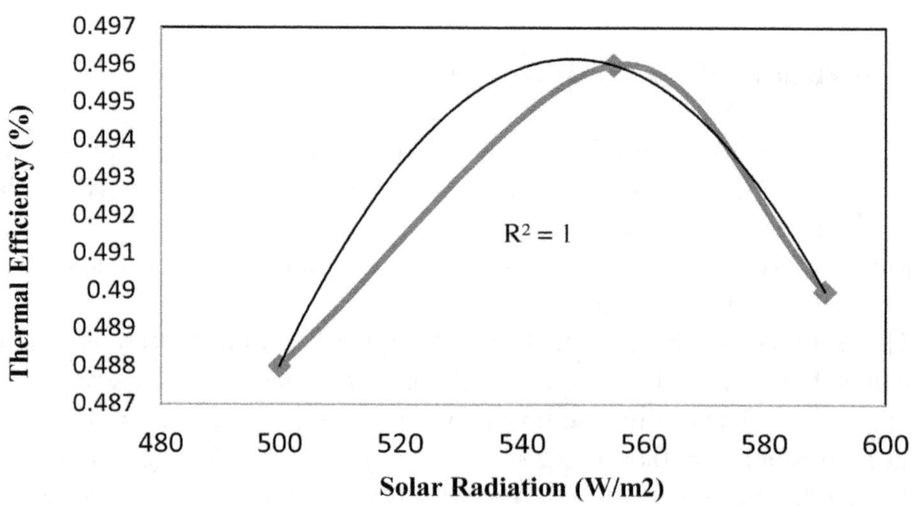

Figure 11. Variation of PTC thermal efficiency with mass flow rate

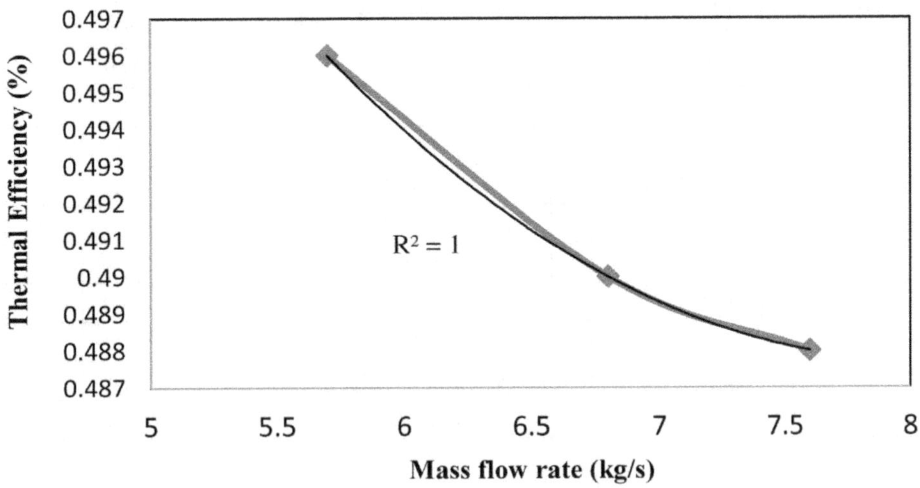

Analysis of Variance Analysis

Further, Analysis of Variance (ANOVA) with 95% of confidence interval is used to determine the significance and adequacy of the developed

Figure 12. Plot of PTC system efficiency with reduced temperature difference (with twisted inserts)

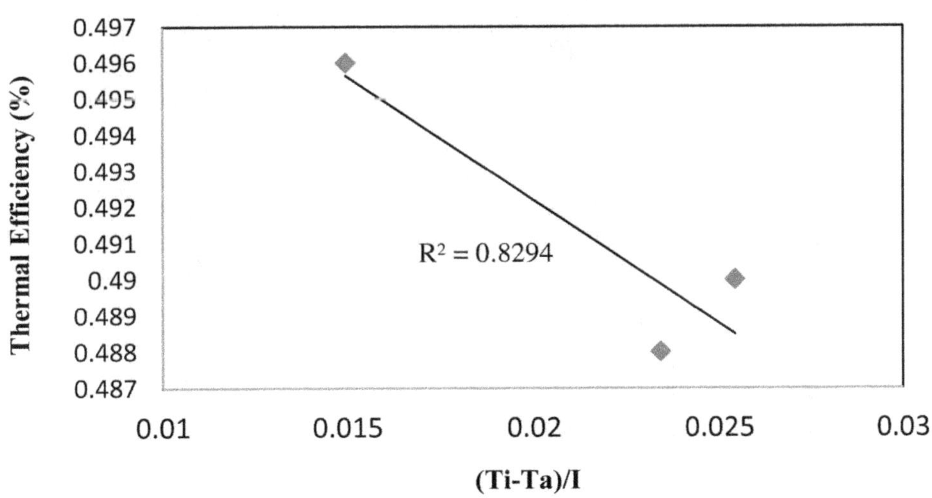

Table 4. Ranking of PSI values of PTC

Trial. No	PSI values	Rank	Trial. No	PSI values	Rank	Trial. No	PSI values	Rank
1	0.3623	10	11	0.3819	2	21	0.3721	4
2	0.3487	16	12	0.3754	3	22	0.3506	15
3	0.3361	23	13	0.3480	17	23	0.3565	12
4	0.3266	25	14	0.3415	19	24	0.3708	5
5	0.3175	26	15	0.3359	24	25	0.3421	18
6	0.3125	27	16	0.3385	20	26	0.3370	22
7	0.3091	28	17	0.3658	7	27	0.3372	21
8	0.3000	29	18	0.3655	8	28	0.3521	14
9	0.2956	30	19	0.3634	9	29	0.3540	13
10	0.4011	1	20	0.3673	6	30	0.3602	11

model. In the ANOVA analysis, the Fisher's statistical test (F-Ratio) and significance probability (P-value) followed by coefficient of determination (R^2) and adjusted-R^2 are determined to test the sufficiency and fitness of

Figure 13. Ranking of PSI values of PTC

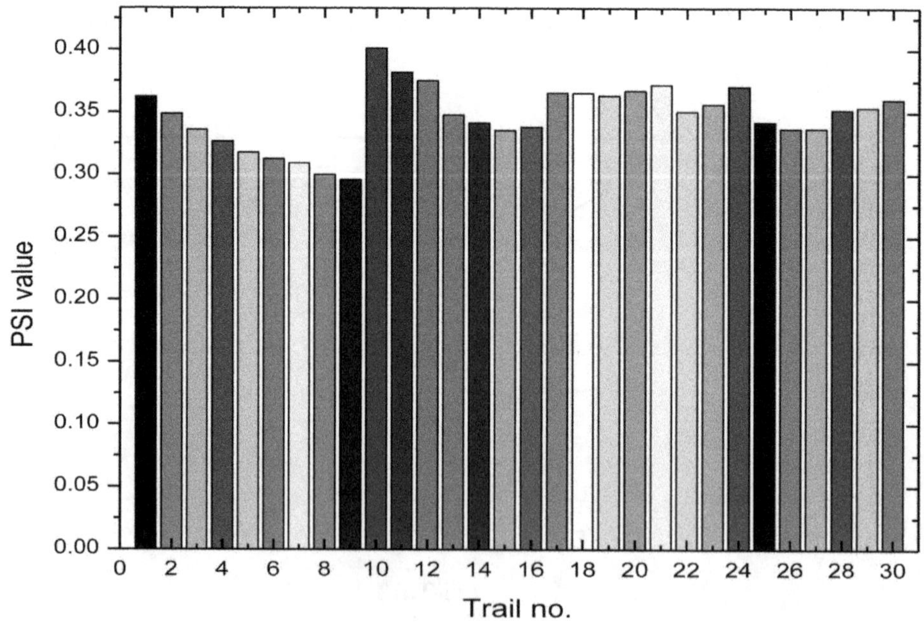

the proposed model. The F-value denotes the most influencing parameters on the performance characteristics while P-value to show the statistical significance of process variables of PTC system. During the analysis, process variables which has larger F-values and lower P-values ($P<0.05$) indicate their significance statistically (Kalogirou, 2020; Kulekci et al., 2011). The R^2 signifies the total variation in the response expected by the model. The purpose of adjusted-R^2 is to analyze the model adequacy and fitness. In this research stepwise elimination method is used to remove the insignificant parameters to adjust the fitted quadratic model (Rao, 2006; Kalogirou, 2020; Kulekci et al., 2011). In ANOVA analysis results are given in Table 5.

Table 5. ANOVA for PSI values of PTC system

Source	DF	Adj SS	Adj MS	F-Value	P-Value
Regression	2	0.006835	0.003418	8.71	0.001
Inlet Temperature (T_i)	1	0.003781	0.003781	9.63	0.004
Solar Radiation (I)	1	0.005054	0.005054	12.87	0.001
Error	27	0.0106	0.000393		
Total	29	0.017436			
R^2=99.97%, R^2(adj)=99.14%, R^2 (Pred)= 98.77%					

The ANOVA for PSI values result shows that the parameter summer height of the receiver (h_s) and winter height of the receiver (h_w) are most significant factor influencing the assessment of PSI values of PTC performance followed by inlet temperature (T_i) in °C and solar radiation (I) are considered as statistically significant. Subsequently, the P values for inlet temperature (T_i) in °C and solar radiation (I) are found to be less than 0.05, which indicate that the model and associated terms are statistically significant (Rao, 2006; Kalogirou, 2020; Kulekci et al., 2011; Rao, & Kalyankar, 2014; Konig, et al., 1985). The other parameters except parameter highlighted in Table 4 are considered to be statistically insignificant. The values of R^2 is calculated to be 0.9997 for PSI values, it implies that 95% of experimental data was well-suited. Higher the R^2 coefficient gives the satisfactory model to the experimental data. Finally, the adequacy and fitness of the model is calculated by adjusted-R^2 values. The high value of adjusted-R^2 (0.9977 for PSI values) supports a high correlation between the experimental and the predicted values (Konig, et al., 1985; and Kumar, et al., 2014).

Confirmatory Test

Optimal setting obtained using the proposed method is verified through confirmatory tests. The optimal setting i.e. T_i= 34.1°C, I=535.60 W/m² at a constant m = 0.0031 kg/s; and A = 0.319 m² is used for confirmatory experiments and the corresponding results are shown in Table 6. The results show that confirmatory tests results are comparable and acceptable with experimental results for the optimal setting.

Table 6. Confirmatory tests for SFPC thermal system

Input Conditions	Output Parameters	Experimental results	Confirmatory results
Test No-10: T_i= 34.1°C, I=535.60 W/m² at a constant m = 0.0031 kg/s; and A = 0.319 m²	T_o in °C	46.3	45.2
	H in W	312.03	309.2
	η_{th} in %	29.4	29.0

SUMMARY

This chapter presented the modeling and optimization of PTC using an integrated MCDM method i.e. entropy with GTMA. Here, the entropy method is used for extraction of priority weights of PTC parameters while GTMA for optimization of PTC parameters. Experiments (30 nos.) have been performed on PTC under different operating/input conditions. The optimization result indicates that trial no. 10 produces the highest PSI value compared to the other trials. The optimal setting is obtained as T_i= 34.1°C, I=535.60 W/m², at constant m = 0.0031 kg/s; and A = 0.319 m², which gives optimal performance parameter i.e. T_o= 46.3 °C, H= 312.03 W and, η_{th} = 29.4% which directly or indirectly increases thermal efficiency (η_{th}) as well as performance of PTC. Additionally, ANOVA analysis is carried out to determine the significance and adequacy of the developed model. The results show that, inlet temperature (T_i) and solar radiation (I) are the major parameters for PTC system and also model fitted data are well fitted with experimental result within the 95%

confidence level. At last, confirmatory tests show comparable and acceptable results with experimental values. As the proposed MCDM technique gives optimized values of the considered operating parameters and efficiency of the collector, and as these are obtained with good accuracy and with statistical significance, the obtained optimized values can be practically implemented for the present solar collector. Hence, it is concluded that the performance of PTC under the considered environment is found to be feasible and the proposed hybrid MCDM method provides acceptable results for PTC.

REFERENCES

Bornatico, R., Pfeiffer, M., Witzig, A., & Guzzella, L. (2012). Optimal sizing of a solar thermal building installation using particle swarm optimization. *Energy*, *41*(1), 31–37. doi:10.1016/j.energy.2011.05.026

Ebrahimi, M., & Rahmani, D. (2018). A five-dimensional approach to sustainability for prioritizing energy production systems using a revised GRA method: A case study. *Renewable Energy*, *135*, 345–354. doi:10.1016/j.renene.2018.12.008

Kalogirou, S. A. (2004). Optimization of solar systems using artificial neural-networks and genetic algorithms. *Applied Energy*, *77*(4), 383–405. doi:10.1016/S0306-2619(03)00153-3

Kalogirou, S. A. (n.d.). *Solar Energy Engineering-Processes and Systems* (2nd ed.). Academic Press.

Konig, W., Wulf, C., & Willerscheid, H. (1985). Machining of fibre reinforced plastics. *CIRP Annals Manufacturing Technology*, *34*, 537-548.

Kulekci, M. K., Esme, U., Er, O., & Kazancoglu, Y. (2011). Modeling and prediction of weld shear strength in friction stir spot welding using design of experiments and neural network. *Materialwissenschaft und Werkstofftechnik*, *42*(11), 990–995. doi:10.1002/mawe.201100781

Kumar, R., Kumar, K., & Bhowmik, S. (2014). Optimization of mechanical properties of epoxy based wood dust reinforced green composite using taguchi method. *International Conference on Advances in Manufacturing and Materials Engineering*, *5*, 688-696. 10.1016/j.mspro.2014.07.316

Lupfert, E., Geyer, M., Zentrum, D., Schiel, W., Esteban, A., & Osuna, R. (2001). Eurotrough design issues and prototype testing at Psa. *Sol Forum*, 1–5.

Mehrabi, D., Mohammadzaheri, M., Firoozfar, A., & Emadi, M. (2017). A fuzzy virtual temperature sensor for an irradiative enclosure. *Journal of Mechanical Science and Technology*, *31*(10), 4989–4994. doi:10.100712206-017-0947-x

Mohanty, S., Rout, A., Patra, P. K., & Sahoo, S. S. (2017). ANFIS based solar radiation data forecasting for energy and economic Study of solar water heaters in eastern India. *International Journal of Control Theory and Applications*, *10*, 179–190.

Rao, R. V. (2006). A material selection model using graph theory and matrix approach. *Materials Science and Engineering*, *431*(1-2), 248–255. doi:10.1016/j.msea.2006.06.006

Rao, R. V., & Kalyankar, V. D. (2014). Optimization of modern machining processes using advanced optimization techniques-a review. *International Journal of Advanced Manufacturing Technology*, *73*(5-8), 1159–1188. doi:10.100700170-014-5894-4

Rao, R. V., & Padmanabhan, K. K. (2007). Rapid Prototyping process selection using graph theory and matrix method. *Journals of Material processing. Technology (Elmsford, N.Y.)*, *194*, 81–88.

Roy B., Shovan S., Balmiki R.D., Das S., Kataky K.P., & Biswas A. (2016). Parametric study of parabolic trough collector- a case study for Silchar, India. *ISESCO Journal of Science and Technology*, *12*(21), 24-29.

Spencer, L. C. (1989). A comprehensive review of small solar-powered heat engines: Part I. A history of solar-powered devices up to 1950. *Solar Energy*, *43*(4), 191–196. doi:10.1016/0038-092X(89)90019-4

Chapter 6
Modeling and Optimization of Solar PV System With Pumped Hydro Energy Storage System

ABSTRACT

Implementation of modified AHP coupled with MOORA methods for modeling and optimization of solar photovoltaic (PV)-pumped hydro energy storage (PHS) system parameter is presented in this chapter. Work optimized the parameters, namely unmet energy (UE), size of PV-panel, and volume of upper reservoir (UR), to get economic cost of energy (COE) and excess energy (EE). The trail no.11 produces the highest assessment values compared to the other trails and provides EE as 16.19% and COE as 0.59 \$/kWh for PV-PHS. ANOVA and parametric study is also performed to determine the significance of the parameters for PV-PHS performance. Investigation results indicate the effectiveness and significant potential for modeling and optimization of PV-PHS system and other solar energy systems.

INTRODUCTION

Standalone SES with feasible energy storage unit can work as an effective micro-grid system with regards to feasibility, proximity to user end requirements, economics and eco-friendliness for satisfying the solar energy needs of people. Design of micro-grid for the user end requirements is an important consideration when the user's load situation, reliability/autonomy requirements

DOI: 10.4018/978-1-7998-3523-3.ch006

are of different scales. The optimization in such situations could be a crucial design aspect so as to make the SES reliable and cost-effective. In the available literature (Hoppmann, et al. 2014; Nair & Garimella, 2010; Mahmoud, 2014; Shaahid, & Elhadidy, 2007) Technical and economic assessment of grid independent hybrid photovoltaic-diesel-battery power system for commercial load in desert environments, Renewable and Sustainable Energy Review, 11: 1794-1810 battery storages have been successfully integrated to SES due to their simplicity and high load stability nature. Another popular storage unit often integrated to SES is PHS technology, which is one of the most mature technologies due to long term energy storage (Saini, 2011; George, 2002). Though in some real projects (Manolakos, et al., 2004; Li, et al., 2012; Ma, et al., 2014), the performances of PHS have been optimized for renewable energy power supply, but the optimum solution might sometimes result in a too large upper reservoir capacity to accommodate (Ma, et al., 2014), and therefore such designs may not be always feasible when the user's load demand is low. Additionally, the feasibility and economic analysis of PHS combined with battery based system has also been studied in literature (Saaty, 1989; Brauers, & Zavadskas, 2006; Karande, & Chakraborty, 2012; Chakraborty, 2011) for some real projects of the world. However, there is a need to optimize sizes of such storage units if these are to be integrated with SES at varied user-load conditions. In the field of sizing under different conflicting criteria of techno-economic costs, reliability, and autonomy for SES, different types of mathematical modelling, optimization methodologies and simulation software have been adopted by various researchers (Sarkar, et al., 2015; Singaravel, et al., 2016; Tzeng, & Huang, 2011; Ma, et al., 2014). SES has also been designed in terms of its optimized sizing for real application projects (Ma, et al., 2014 & Kaldellis, et al., 2013). Different mono-objective and multi-objective optimization functions have been used in the sizing of SES. For example, Genetic Algorithm (GA) gives the flexibility to consider all the aspects of RES design considerations such as technical feasibility, economical feasibility, optimality and reliability. Mathematical equations for major components of the systems are used for sizing and modeling.

The hour-by-hour simulation program has been developed and systems are optimized with respect to their Life Cycle Cost (LCC), COE, and UE using integrated MCDM method. Both mono-objective and double-objective function optimizations have been implemented for each RES. A sensitivity analysis has been performed by considering the effect of percentage UE (from 0% to10%, i.e. reliability between 90% and 100%) on the COE for different storage capacity of PHS, so as to design optimum storage units for the end user requirements. The methodology is finally applied on an academic block of an educational institute at Silchar, India having low load factor user-load condition, and based on the results various conclusions are drawn at the end.

The PHS system with PV power generation can store large amount of energy for long period and working principle is also very simple. Bulk storage, reliability, and maintenance free storage are the reasons why PHS is preferred to battery bank. Although batteries are less costly, they require more maintenance and need to be replaced after 3-4 years of use. Moreover, depth of discharge and state of charge of the battery bank need to be correctly maintained at the set level in case of battery, which need not to be rigorously followed for PHS, hence the latter has higher reliability. Moreover, the number of autonomy days can also be higher in case of PHS. However, as

Figure 1. The stand-alone PV-PHS based SES

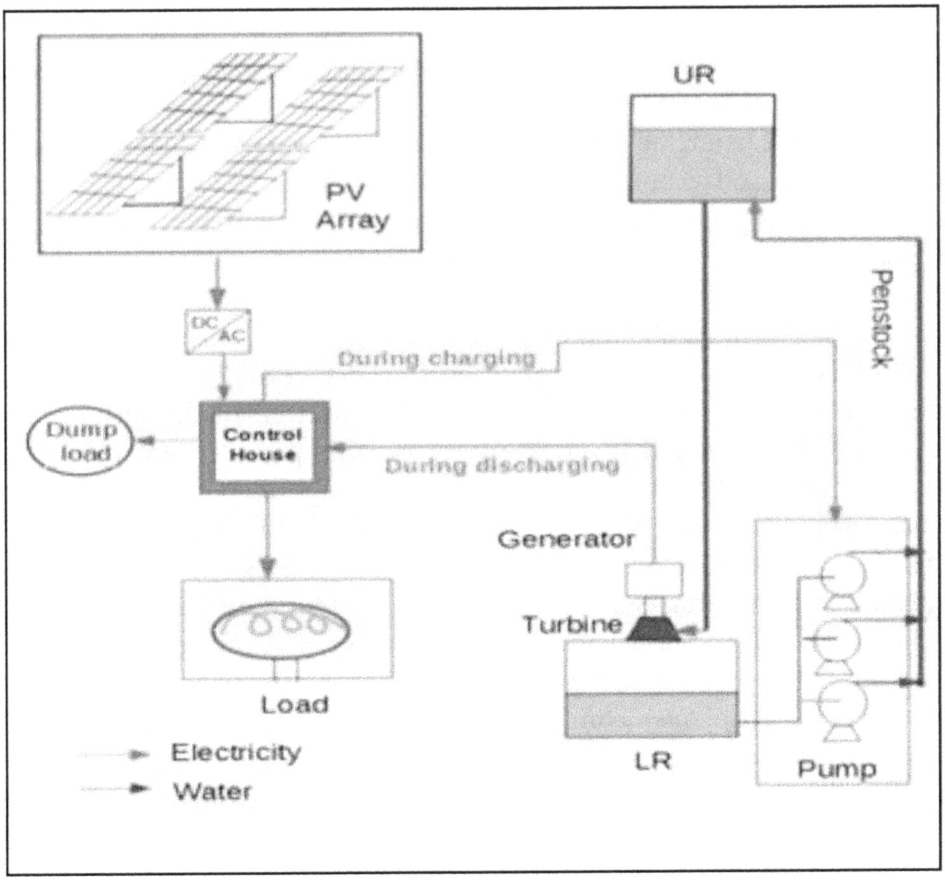

the construction cost of PHS is high, hence it is to be used in case of bulk storage requirements, for e.g. remote village electrification, micro-grid, etc.

Figure 1 demonstrates the operation of the stand-alone PV-PHS based SES to be designed for academic building power consumption. Here major components are namely PV, pump, turbine, upper reservoir (UR), lower reservoir, generator and DC-AC converter. PV power first satisfies the customer load demand then consequently surplus energy goes to the pump units. And pump elevates the water from lower reservoir to upper reservoir, and when energy deficit occurs, the turbine/generator unit serves the load. It can be noted that (Figure 1) double-penstocks have been used- one is for charging and another for discharging of the water; therefore, the charging and discharging could be possible simultaneously. The double-penstock type PHS system has been adopted in several real projects such as (Saaty, 1989; Greacen, 2006 & Jamil., et al., 2012) and this type of configuration is more efficient and highly reliable to power supply. A number of parallel pumps have been connected together for increasing the flow rate of water for a give head. All the charging, discharging and flow water and currents are control through the controlling house. For taking care of excess energy a dump load is also connected with control center.

On the other hand, the performance of a SES is very much dependent on operating/input and response/output parameters which mainly affect the sizing of the SES thereby requiring a performance optimization. Hence, optimization of SES parameters is essential. Although many investigators have worked on optimization of solar systems (Nair & Garimella, 2010; Mahmoud, 2014; Shaahid, & Elhadidy, 2007; Saini, 2011; George, 2002), they are limited to single objective optimization. In addition, importance of criteria weights and effects of both beneficial and non-beneficial parameters on the performance of SES have been ignored. In order to overcome this, present chapter proposes a hybrid integrated MCDM method, which consists of AHP with MOORA (Manolakos, et al., 2004; Li, et al., 2012; Ma, et al., 2014) method. Here, AHP is used for extraction of precise priority weights while MOORA for optimal selection of SES parameters. Also, none of the literature on modeling and optimization of SES parameters using AHP-MOORA method is available. In addition, the proposed method offers various merits like ability to solve multi-criteria problems, considers multiple criteria with different data measurements, considers subjective and objective data during optimization, easy to maintain consistency, less computational time required, very simple and stable technique, minimum mathematical calculation involved, which may be quite useful and helpful to the decision makers who may not have a strong background in mathematics.

EXPERIMENTAL DETAILS

Systems Modeling

The mathematical equation for individual major components of the renewable energy systems is proposed in this section. An hour by hour simulation program is then developed to size the PV panel, battery banks, UR, inverter, pump and turbines.

PV Array Modeling

The detailed manufacturing data of the PV panel are provided in Table 1. The equivalent circuit of PV cell is shown in Figure 2. The power produced from the PV panel can be also directly evaluated by using Equation (1).

$$P_{PV}(t) = N_{PV}Y_{PV}\left(\frac{G(t)}{G_O}\right)\left[1-\alpha(T_C - T_O\right] \quad (1)$$

If the effect of temperature is neglected, then Equation (1) can be expressed as-

$$P_{PV}(t) = N_{PV}Y_{PV}\left(\frac{G(t)}{G_O}\right) \quad (2)$$

Efficiency of the PV panel can also be used for obtaining solar power from PV panel as follows Rawat, & Chandel, 2013):

$$P_{PV}(t) = \eta P_V \bullet A \bullet G(t) \quad (3)$$

So, the energy generated through solar PV panel per day can be determine by using Equation (4)

$$E_{PV} = \int_{0}^{24} P_{PV}(t)dt \tag{4}$$

where $P_{PV}(t)$ is the net power produced by PV panel (kW), N_{PV} is the number of PV panels, Y_{PV} is the PV derating factor, accounting for the factors like ageing, soiling, wiring losses, shading, and so on; $G(t)$ is global solar insolation at any time 't', G_o is the standard solar insolation (1kW/m^2), α is the negative temperature coefficient of PV module, TC is solar cell temperature and T0 is the standard cell temperature (25o^C), ηPV $_i$s efficiency of PV panel, A is area of PV module (m2)· All the constant parameters that have been used here are provided in Table 2. International Electrochemical Commission (IEC) has developed and published a number of module/photocell and component measurement and qualification standards (Wohlgemuth, 2012). Belenov et al. (2016) examined solar PV based power plant operation on the roof of an administrative building in Russia. Hourly load data were gathered, and the power plant produced over 60 kWh during the sunny days. They investigated in detail the time characteristics of the largest hourly power output of the plant in the daytime for various seasons, irrespective of the cloud thickness. Within two years of observations, approximately 1100 hours of energy consumption were reported and transmitted by the electricity network to the internal system of the building. The effect of the installation angle of the solar panels on electricity generation at different solar radiations was analysed. In order to improve the cell efficiency of the PV module, heat absorbed by the PV panel can be taken away by integrating a cogeneration technology with a cooling system. Nesterenkov & Kharchenko, (2018) designed a power plant for farmers by integrating cogeneration technology with low solar radiation compatible silicon photo/solar cell. The cell efficiency is improved with combining photocells at a relatively high temperature and liquid cooling, which allow high-potential heat to be obtained and transferred to the heat transport system in the counter-flow mode of the coolant. Panchenko et al. (2018) developed three-dimensional photovoltaic solar thermal module (a cogeneration system of electric power and thermal energy) using Ansys software for simulation and thermal performance measurement of the hybrid module. This work is a testimony of the usefulness of solar PV system that it can generate solar thermal energy as a secondary output in addition to its primary output of electric power. Panchenko et al. (2018 & 2020) again performed simulations of solar photovoltaic thermal modules for visualizing thermal processes, coolant velocity and flow lines of coolant in the three-dimensional modules.

Pumping System Modeling

The quantity of water pumped per second to the upper reservoir can be determined from Equation (5) as-

Figure 2. The equivalent circuit of PV cell

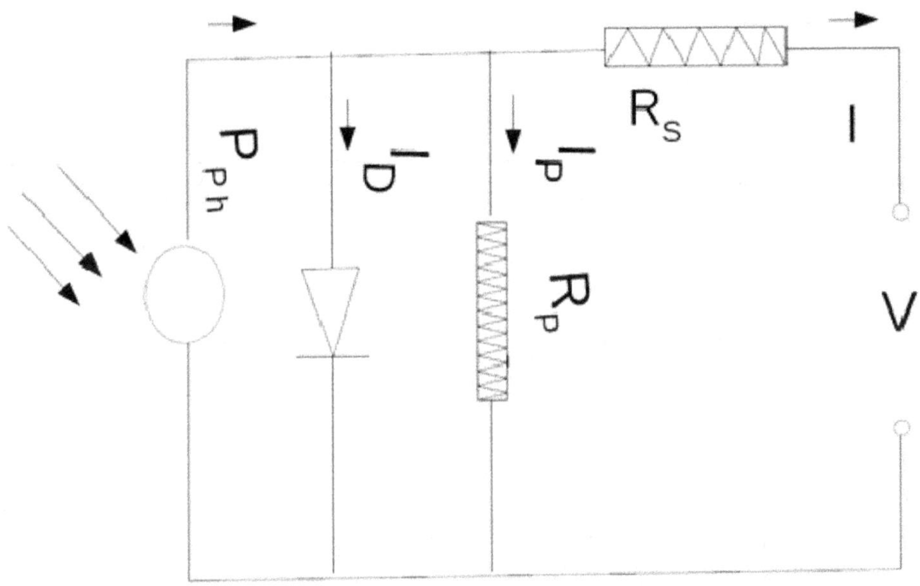

Table 1. Specification of Heckert NeMo@54p PV panel

Characteristic	Valve
Type	Polycrystalline(54 cell)
Open circuit voltage(V_{oc})	33.3V
Optimum operating voltage (Vmp)	27.3V
Short circuit current (I_{sc})	8.67A
Optimum operating current (Imp)	7. 98A
Maximum power at STC(P_{max})	200Wp
efficiency of module(η_{PV})	14.6%
Area of module (A_{PV})	1.4×.999 m^2
Temperature coefficient of I_{sc}	0.05
Temperature coefficient of V_{oc}	-0.32
Temperature coefficient of power α	-0.42

Table 2. A summary of constant parameters

Parameters	Value
Inverter efficiency (ηe_{ff})	85%
Pumping efficiency (ηp)	80%
Leakage and evaporation loss factor (α)	0.05
Turbine efficiency (ηT)	70%
Derating factor (Y_{PV})	85%
Total head (h)	30m
Nominal interest (i')	7%
Annual interest (f)	4%
Life of project (n)	25year
Coefficient of self-discharge (γ)	0.002
Days of autonomy (ηd_{ay})	1 day

$$q_p(t) = \frac{\eta_p \cdot P_{Pump}(t)}{\rho \cdot g \cdot h} = C_P \cdot P_{Pump}(t) \quad (5)$$

where P_{Pump} is power supplied by the PV panel to the pump (kW), h is the total head (m), g is the acceleration due to gravity (9.8 m/s^2), ρ is standard water density (1000 kg/m3); ηp is the total net pumping efficiency, and Cp is the water pumping coefficient of the pump unit (m3/kWs); So, total amount of water pumped in an hour is given by Equation (6)

$$QP(t) = qp(t) \times 3600 \quad (6)$$

The total number of parallel pumps needed is determined by Equation (7):

$$N_{Pump} = \frac{P_{\max}}{P_o} \quad (7)$$

where $P_{\max}$ the maximum amount of surplus power coming from PV is array (kW); P_o is rated power of single pump (kW).

Turbine System Modeling

When any type of energy short-fall appears on user-end side, water is drawn from the UR and passed on to the turbine unit. The power generated through the turbine for a given flow rate of water is determined as per Equation (8).

$$P_T(t) = \eta T \bullet \rho \bullet g \cdot h \cdot qT(_t) = CT\,.\,qT(_t) \tag{8}$$

where ηT $_i$s the overall efficiency (including pipe losses) of the turbine, qT($_t$) is the volumetric flow rate of water towards the turbine (m3/^{s}), CT $_i$s the coefficient of turbine. So, total amount of water input into the turbine in an hour is given by Equation (9)

$$QT_(t) = qT_(t) \times 3600 \tag{9}$$

Upper Reservoir (UR) Modeling

The amount of water stored in the UR should be sufficient to satisfy power demand in case if no power could be supplied for several consecutive days (Ma, et al. 2015). during these power cuts, several days of autonomy have to be considered. The amount of gravitational-potential energy stored in the UR can be obtained as-

$$E_{Cap} = \theta \cdot n_{day} \cdot E_{load} = \frac{\eta_T \cdot \rho \cdot V_{UR} \cdot g \cdot h}{3.6 \times 10^6} \tag{10}$$

where E_{Cap} is the energy storage capacity of a water reservoir (kWh); V_{UR} is the volume or storage capacity of the UR (m^3). Therefore, the total quantity of water stored in the UR at any time‘t’ is determined by:

$$Q_{UR}(t) = Q_{UR}(t-1)(1-\alpha) + QP_(t) - QT_(t) \tag{11}$$

where α (equal to 0.05) is the evaporation and leakage loss coefficient. The water quantity of the UR should be subject to the following constraints:

$$Q_{UR\,\min} \le Q_{UR} \le Q_{UR\,\max} \tag{12}$$

where $Q_{UR\max}=V_{UR}$ and $Q_{UR\min}$ is minimum water storage in UR which is usually set as zero. The quantity of water level in the UR can be considered as the state of charge (SOC) in percentage of the storage tank capacity. Therefore, UR's SOC can be determined for time't' as Equation (13)

$$SOC(t)(\%) = \frac{Q_{UR}(t)}{V_{UR}} \times 100 \tag{13}$$

Energy Shearing Model for Electrical Energy Generation

The generalized energy balance model of the solar power generation system at time't' can be expressed as:

$$P_{PV}(t) \bullet \eta i_{nv.} = \mathrm{P}L_{(}t) + \mathrm{P}P_{ump(}t) + \mathrm{P}D_{ump(}t) \tag{14}$$

where $\eta i_{nv.}$ is the inverter efficiency, which is defined as the ratio of the inverter's AC output power to the DC input power, $\mathrm{P}L_{(}t)$ is direct satisfied load through the PV solar power system; $\mathrm{P}P_{ump(}t)$ is amount of surplus power transferred to the pumps in order to charge UR and $\mathrm{P}D_{ump(}t)$ is the amount of excess energy delivered to a dump load for taking care of other components in the system. However the load demand is mainly covered by three sources such as PV, pump and battery, therefore the energy balance model of load consumption is given by:

$$P_{TL}(t) = P_{L}(t) + P_{Turbine}(t) \tag{15}$$

where $P_{L}(t)$ is the power that is directly delivered from the RES generator; $P_{Turbine}(t)$ is the power produced by the turbine generator unit. When the net load is negative or zero, no supplementary energy is required and thus $P_{Turbine}(t)$ is zero.

Cost Analysis for SES' Main Components

Total Life Cycle Cost (LCC) is used to analyze the SES's economic performance. LCC includes initial cost, installation, replacement, and operation and maintenance cost (O and M). The total LCC can be calculated according to Equation (16)

$$LCC = CC + O\,\&\,M + RC \tag{16}$$

Interest Rate

The annual interest rate is the discount rate used to convert between one-time costs and annualized costs. It is related to the nominal interest rate (Konig, et al., 1985) as-

$$i = \frac{i' - f}{1 + f} \tag{17}$$

Where i is the interest rate, i' nominal interest rate, and f is the annual inflation rate as shown in Table 2.

Replacement Cost

Total Replacement Cost (RC) of the main components can be calculated by Equation (18) as-

$$RC = \sum_{i}^{N} [C_{rep_i} \times f_{rep_i} \times SFF_i - S_i \times SFF] \tag{18}$$

f_{rep_i}, a factor arising because the component lifetime can be different from the project lifetime, which is defined in Equation (19)

$$f_{rep_i} = \frac{CRF}{CRF_i} \tag{19}$$

SFF= Sinking Fund Factor is defined as Equation (20)

$$SFF = \frac{i}{(1+i)^{n-1}} \tag{20}$$

Then selvage value of the components is defined as Equation (21)

$$S = C_{rep} \times \frac{L_{rem}}{L_{comp}} \tag{21}$$

L_{rem} is remaining life of the components, L_{comp} is total life of components and C_{rep} is the replacement cost factor.

CRF is Capital Recovery Factor, which is defined as in Equation (22)

$$CRF = \frac{i(1+i)^n}{(1+i)^n - 1} \tag{22}$$

So, the total Annualized Life Cycle Cost (ALCC) can be expressed as-

$$ALCC = LCC \times CRF \tag{23}$$

where n is the total life of project.

Cost Information of the Components

The cost of major components for all the components namely PV module, Pump, Turbine, and Reservoir can be obtained for the location where SES is designed. The capital cost, operation and maintenance cost and replacement cost of all the components have been considered. The cost information for major components and their corresponding lifespan are summarized in Table 3.

Table 3. A summary of cost and lifecycle of system's major components

Component	Unit Capital Cost($)	Life cycle (Year)	Replacement cost ($)	Operation and Maintenance cost $/kW-year in % of capital cost
PV panel (200Wp)	300	25	200	0.2%
Inverter	900/kW	15	900	0.5%
Pump	230 /kW	10	180	0.5%
Turbine with pipes	600/kW	10	400	0.5%
Upper reservoir	150/m^3	35	150	0.5%

Levelized Cost of Energy (COE)

The levelized COE is considered as a principal cost of economics for figuring out merit of the all systems. It is the techno-economic criteria for the sizing of the SES. The optimized performance of the SES depends on the lowest value of COE under the constraint criterias of unmet energy and excess energy. COE is ratio of the annualize life cycle cost of the system to the yearly energy demand. In other words, it is the cost of per unit electricity $/kWh. The sizing of the SES should be such so as to minimise COE

$$COE = \frac{ALCC}{E_{Load}} \tag{24}$$

System Reliability Index

In this present study, the system reliability index is evaluated based on the percentage of unmet energy (% UE) i.e. percentage of unmet load per year, which is defined as the summation of total power supply failure to the total energy demand over a year (Kulekci, et al., 2011). The reliability study is done to evaluate whether a system is able to fulfill the load demand, and if there is any deficiency, then to calculate the percentage of unmet energy. Another most widely used important reliability index is percentage of excess energy (EE) in a year. The indices UE and EE are calculated by Equations (25 &26) as follows-

$$UE(\%) = \frac{\sum_{1}^{8760} P_{TL}(t) > [P_L(t) + P_{Turbine}(t)]}{E_{Load}} \times 100 \tag{25}$$

$$EE(\%) = \frac{\sum_{1}^{8760} P_{Dump}(t)}{\sum_{1}^{8760} P_{PV}(t)} \times 100 \tag{26}$$

Integrated AHP-MOORA Method

An integrated MCDM method consists of an AHP (Saaty, 1989) combined with MOORA (Brauers, and Zavadskas, 2006; Karande, & Chakraborty, 2012; Chakraborty, 2011; Sarkar, et al., 2015; Singaravel, et al., 2016) which is presented for optimization of the performance of the Solar Energy System with an energy storage system (PHS), which is applied to fulfil the power consumption demand of an academic building. Various input and output parameters obtained from the performance of the solar energy system are considered in the model for obtaining the optimal values of the operating parameters to fulfill the hourly power demand of the academic building. The schematic diagram of AHP-MOORA method is shown in Figure 3 and the detailed steps of AHP-MOORA based decision-making method are as follows:

Step 1: Development of decision matrix

Optimization process starts with the decision matrix, which includes a number of response parameters and corresponding process parameters. First, formulation of decision matrix is done based on a number of process/response parameters of the system. Various designs of experimental techniques are used for the formulation of the decision matrix. In the decision matrix, performance values for each of the response parameters with respect to the process parameters are listed. The decision matrix is formulated using the following expression.

$$D_{ij} = \begin{bmatrix} D_{11} & D_{12} & \cdots\cdots & D_{1q} \\ D_{21} & D_{22} & \cdots\cdots & D_{2q} \\ \cdots & \cdots & \cdots\cdots & \cdots \\ D_{p1} & D_{p2} & \cdots\cdots & D_{pq} \end{bmatrix} \text{For } (i=1,2,\ldots p,\ j=1,2,\ldots q) \tag{27}$$

where D_{ij} represents the response value of ith responses on jth process parameter, p represents the number of experiments and q represents the number of response parameters.

Step 2: Formulation of pairwise comparison decision matrix and computation of precise priority weights of the response parameters using AHP

Figure 3. Flow chart of AHP-MOORA method

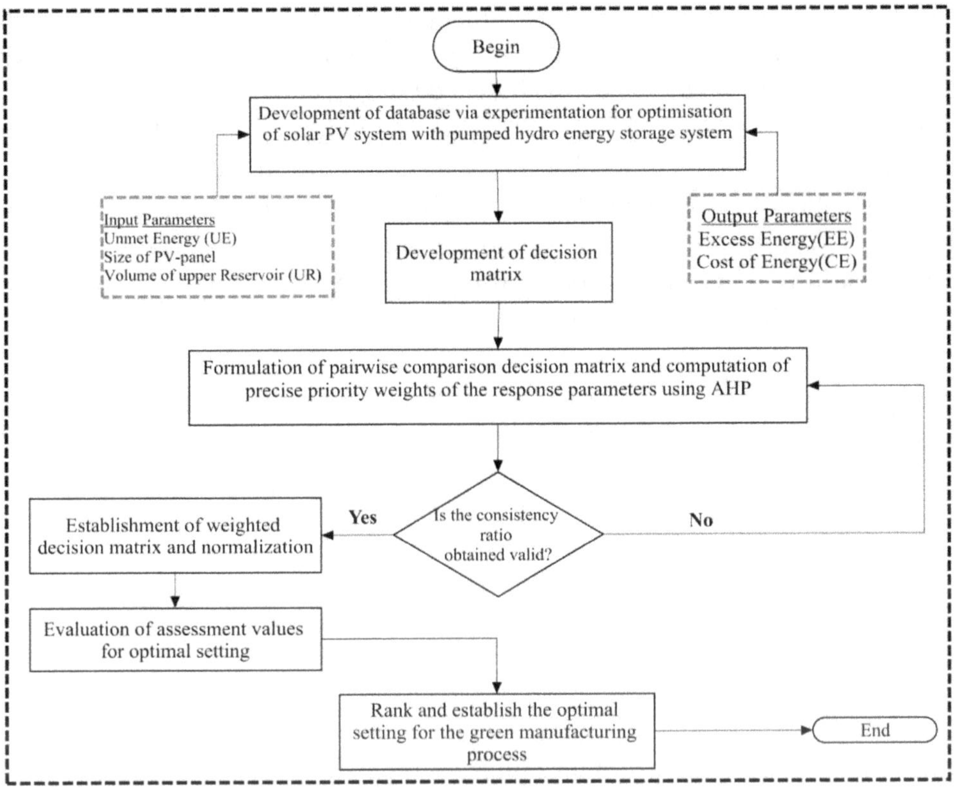

In order to find out the priority weights of the response parameters, the pairwise comparison matrix is required. Here, precise priority weights for each of the response parameters are determined using AHP method (Saaty, 1989). Initially, pairwise comparison decision matrix between each of the response parameters is developed. Simultaneously, the precise priority weights for each of the response parameters are determined using the fundamental scale of the AHP process with the *i*th response and *j*th criteria of the response parameters using the following expression (Tzeng, & Huang, 2011).

$$C_{ij} = \begin{bmatrix} C_{11} & C_{12} & \dots\dots\dots\dots & C_{1j} \\ C_{21} & C_{22} & \dots\dots\dots\dots & C_{2j} \\ \dots\dots & \dots\dots & \dots\dots\dots\dots & \dots\dots \\ C_{i1} & C_{i2} & \dots\dots\dots\dots & C_{ij} \end{bmatrix} C_{ij}=1,\ C_{ij} = \frac{1}{C_{ji}},\ C_{ji} \neq 0 \quad (28)$$

where, $C_1, C_2...C_j$ denotes the response parameters, C_{ij} is the relative importance of *i*th response parameter with *j*th response parameter. The relative importance of two response parameters is rated using the fundamental scale of AHP with the digits 1, 3, 5, 7 and 9. In this digit 1 denotes "equally important", 3 for "slightly more important", 5 for "strongly more important", 7 for "demonstrably more important" and 9 for "absolutely more important" and 2, 4, 6, 8 for "slightly differing judgment" (Tzeng, & Huang, 2011). Thereafter, priority weights (ϕj) for each response parameter are determined as a row average of the resulting pairwise comparison matrix using Equations (29) and (30) (Brauers, and Zavadskas, 2006).

$$_{Mj} = \left[\prod_{j=1}^{n} C_{ij} \right]^{1/n} \quad \text{For } i,j=1,2,\ldots n \tag{29}$$

$$\varphi_j = \frac{_{Mj}}{\sum_{j=1}^{n} {_{Mj}}} \tag{30}$$

where M_j represents the geometric mean of the pairwise decision matrix (C_{ij}) for *j*th response parameter, ϕ_j is the priority weight of *j*th response parameter, and *n* denotes the number of response parameters.

After that, consistency of the pairwise comparison decision matrix and precision of the priority weights (Karande, & Chakraborty, 2012) obtained from AHP method are verified by Consistency Index (*I*) and Consistency Ratio (*R*) (Brauers, and Zavadskas, 2006):

$$I = \frac{(\lambda_{Max} - n)}{n-1} \tag{31}$$

$$R = \frac{I}{R_I} \tag{32}$$

where λ_{Max} represents the largest eigenvalues of the pairwise comparison decision matrix (C_{ij}), *n* represents the number of experimental runs considered in this study, R_I denotes the random index and it depends on several response parameters.

Step 3: Establishment of weighted decision matrix and its normalization

In this step, weighted decision matrix using Equation (33) is developed. It involves the list of priority weights for each of the response parameters obtained by the AHP method with respect to the different experimental runs. This measures the performance values for each of *i*th experimental runs with respect to the *j*th response parameters. After that conversion of different units of the performance values in the decision matrix into comparable unit of performance values is done in the normalization process. The normalized decision matrix can be calculated using the following expression:

$$WD_{ij} = \begin{bmatrix} \varphi_1 \times D_{11} & \varphi_1 \times D_{12} & \cdots\cdots\cdots & \varphi_j \times D_{1j} \\ \varphi_2 \times D_{21} & \varphi_2 \times D_{22} & \cdots\cdots\cdots & \varphi_j \times D_{2j} \\ \cdots\cdots & \cdots\cdots & \cdots\cdots\cdots & \cdots\cdots \\ \varphi_j \times D_{i1} & \varphi_j \times D_{i2} & \cdots\cdots\cdots & \varphi_j \times D_{ij} \end{bmatrix} \tag{33}$$

$$WD_{ij}^{*} = \frac{WD_{ij}}{\sum_{i=1}^{n} WD_{ij}} \tag{34}$$

where, WD_{ij}^{*} represent the performance weights of *i*th experimental runs on *j*th response parameters and *n* represents the number of response parameters. WD_{ij} represents a dimensionless number in the range [0, 1] representing the normalized performance weights of *i*th experimental runs on *j*th response parameters.

Step 4: Evaluation of Assessment Values for optimal setting

Next, Assessment Values (AV) for each of the response parameters are evaluated in order to assess the optimal selection of process parameters. The evaluation of AV for *i*th experimental runs is done to convert the multi-attribute problem into a single attribute problem subjected to certain constraints, with consideration of both beneficial and non-beneficial parameters using the following expression (Brauers, and Zavadskas, 2006; Karande, & Chakraborty, 2012; Chakraborty, 2011).

$$AV_i = \sum_{j=1}^{g} WD_{ij}^* - \sum_{j=g+1}^{n} WD_{ij}^* \quad (35)$$

where g represents the number of response parameters to be maximized, $(n\text{-}g)$ represents \the number of response parameters to be minimized, AV_i represents the assessment value of i*th* experimental setting with respect to all j*th* response parameters, and ϕ_j represent the precise priority weights of the *j*th response parameters.

Step 5: Rank and establish the optimal setting

At last, optimal selection of process parameters by ranking the experimental runs based on AV using Equation (35) is done. The experimental run with maximum assessed value is optimal among all experimental runs. The optimal setting or optimal combination of process parameters provides the optimal responses.

Modeling of SES System

In this section, the modeling of SES is presented using an integrated method i.e. AHP with MOORA. During modeling, first the development of decision matrix is done based on the data of SES by varying the operating parameters like UE in %, Size of PV-panel, and Volume of UR in m^3 as shown in Table 4. Second, the precise priority weights of the response parameters such as EE in %, and COE in $/kWh are determined using the AHP method (Saaty,

Table 4. Pairwise comparison matrix using AHP

Response parameters	Excess Energy(EE)	Cost of Energy(COE)
Excess Energy(EE)	1	0.2
Cost of Energy(CE)	5	1

1989). Here, first pairwise comparison matrix between the each of the response parameters is formulated using Equation (35) and subsequently, precise priority weights for each of the response parameters of SES process are determined using the Equations (36) and (37). The pairwise comparison matrix between response parameters is depicted in Table 4.

After that, consistency of the decision matrix and precision of the priority weights (Karande, & Chakraborty, 2012) are verified through consistency index (I) and consistency ratio (R) (Eqs. 6.38-6.39). The results of the precise priority weights of the SES response parameters are tabulated in Table 5.

Table 5. Priority weights of the SES response parameters

Priority Weights(ϕ)	Response parameters	
	Excess Energy (EE)	Cost of Energy (COE)
	0.7758	0.2242

The precise priority weights (ϕ) for the SES response parameters such as EE and COE are calculated to be 0.7758, and 0.2242, respectively. The result indicates that response parameter EE is the most influencing parameter followed by COE. Also, Consistency Index and Consistency Ratio for SES response parameters are calculated to be 0.00017 and 0.00029, which are less than 10%. This indicates that the priority weights of the response parameters for the SES and the corresponding pairwise decision matrix shows higher accuracy and better consistency. Thereafter, formulation of weighted decision matrix using Equation (40) is done, in which priority weights for each of the SES response parameters are obtained by an AHP method with respect to the different experimental runs conducted. This measure the performance values for each of the response parameters with respect to each of the experimental trail nos., and the results of the weighted decision matrix are tabulated in Table 6.

After that, normalization of performance values of the weighted decision matrix for each of the SES response parameters is performed to convert the original sequence data into comparable sequence values using Equation (41). The normalization result using MOORA method is depicted in Table 7.

At last, *AV* for each of the response parameters is computed using Equation (42). This converts the multi-criteria problem into a single criteria problem

Table 6. Weighted decision matrix of SES process

Trail. No	Output parameters	
	Excess Energy(EE)	Cost of Energy(CE)
1	0.1174	0.0288
2	0.0714	0.0278
3	0.0641	0.0270
4	0.0555	0.0272
5	0.0668	0.0213
6	0.0874	0.0141
7	0.0769	0.0142
8	0.0617	0.0164
9	0.0488	0.0226
10	0.0640	0.0129
11	0.0619	0.0119
12	0.0460	0.0207

subjected to certain constraints, with consideration of both beneficial and non-beneficial parameters (Karande, & Chakraborty, 2012; Chakraborty, 2011; Sarkar, et al., 2015; Singaravel, et al., 2016). The AV for each of the experimental trails of the SES are tabulated in Table 8.

Table 7. Normalized decision matrix of SES process

Trail. No	Output parameters	
	Excess Energy(EE)	Cost of Energy(CE)
1	0.1513	0.1284
2	0.0920	0.1240
3	0.0827	0.1205
4	0.0715	0.1213
5	0.0861	0.0951
6	0.1127	0.0627
7	0.0991	0.0632
8	0.0796	0.0731
9	0.0629	0.1006
10	0.0825	0.0577
11	0.0798	0.0532
12	0.0593	0.0923

Table 8. Assessment values of SES process

Trail. No	AV
1	0.4958
2	0.3690
3	0.4558
4	0.4134
5	0.4405
6	0.5073
7	0.4553
8	0.3907
9	0.3569
10	0.3845
11	0.7309
12	0.3333

RESULTS AND DISCUSSION

Based on the detailed modeling (both theoretical modeling and experimental modeling) of the components of the solar PV-PHS system, a techno-economic performance (COE and other parameters) of the present system is evaluated and also the same has been optimized. For that a parametric analysis is done to understand the theoretical performance of the system under different operating as well as the cost and reliability parameters for e.g. solar radiation, load condition, components costs, excess or unmet energy, etc.

Parametric Analysis

The proposed SES system is applied in a real technical institute academic block with low load factor, which is situated in National Institute of Technology Silchar, India. This academic block is a 4-storey building, which consists of large rooms for laboratories and faculty cabins. A survey for the electricity consumption of the building was conducted for collecting all load data over a year. Figure 4 shows the hourly power demand for winter with respect to weekday and weekend load variations. And Figure 5 illustrates that average load demand of winter in weekday is 4.055 kW, in the weekend is 1.22 kW, and overall average power demand is 2.633 kW. Similarly, Figure 6 and Figure

Figure 4. Hourly energy demand variation for winter

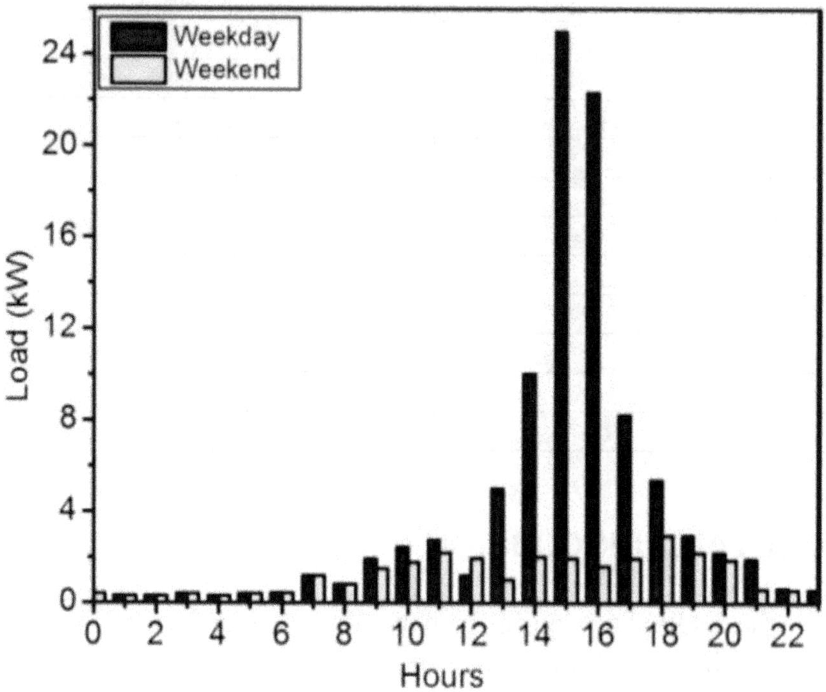

7 demonstrate hourly load demand and average load demand for the summer session (average load demand for weekday of 5.48 kW, average load demand for weekend of 1.96 kW, and overall average energy demand of 3.72 kW).

Figure 8 shows the monthly load demand profile for the academic building. During the summer break, which is in the month of June and July, the laboratory load is zero and therefore, the net load demand is also low. To account for the variation in daily load demand, a random variability factor of 2% for both day-to-day and time–step–to–time profile was considered. Therefore, by considering all variations of the load demand, the overall load factor and daily average electricity demand of the academic block is estimated as 0.131 and 87.8 kWh/day respectively. The load factor is defined as the ratio of the average load to the peak load. For site under consideration i.e. Silchar (latitude 24.833° N and longitude 92.77° E), the solar energy scaled annual average irradiation ($G(t)_{avg}$) is 4.51 kWh/m^2/day, the clearness index of the atmosphere is 0.494 (Singaravel, et al., 2016; & Ma, et al. 2015), Figure 9

Figure 5. Daily average load demand variation for winter (weekday, weekend and overall)

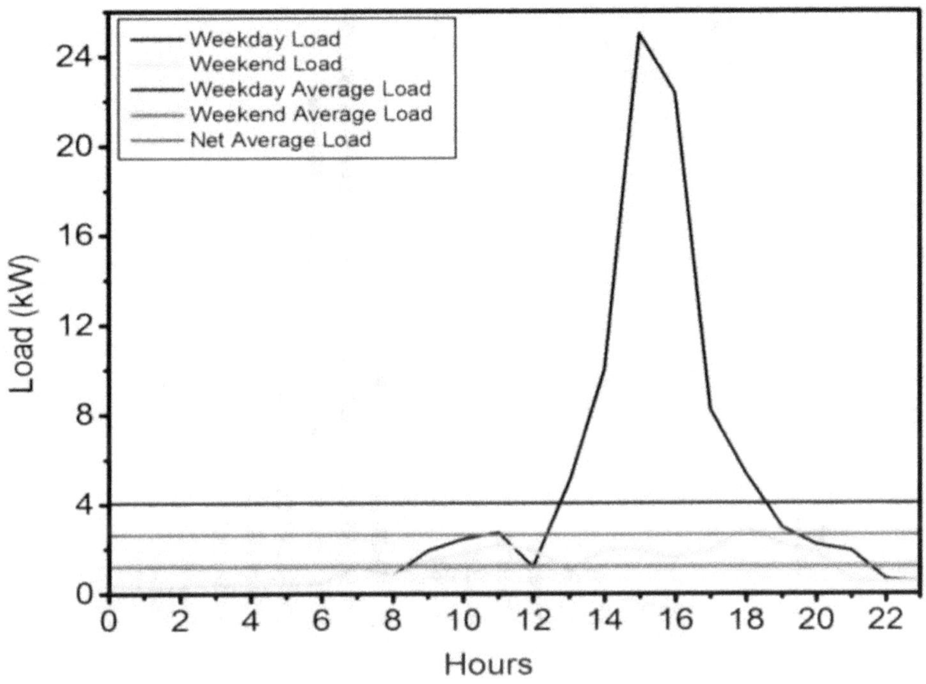

and Figure 10 illustrate the monthly and hourly variation of solar insolation respectively, and wind speed in this region is also unfavourable (Singaravel, et al., 2016).

Table 9 demonstrate the variation of COE with percent UE, which summarizes the variations of the variables- PV panel size, UR volume, and UE. It can be seen that 0.5% of UE shows significant reduction of COE, volume of UR (i.e. 32.2% and 39.2% reduction of COE and UR respectively compared to 0% UE), and capacity of PV plant is also significantly reduced. From Table 4, it can be further observed that 9% UE configuration would be the better design solution as it has least EE (16.18%), smallest capacity PV panels (46.92 kW; 235 pieces of area 328.16m^2), almost lowest COE ($0.588/kWh), and also smaller UR capacity (1188.29m^3) compared to 0% UE configuration.

Figure 6. Hourly power demand variation for summer

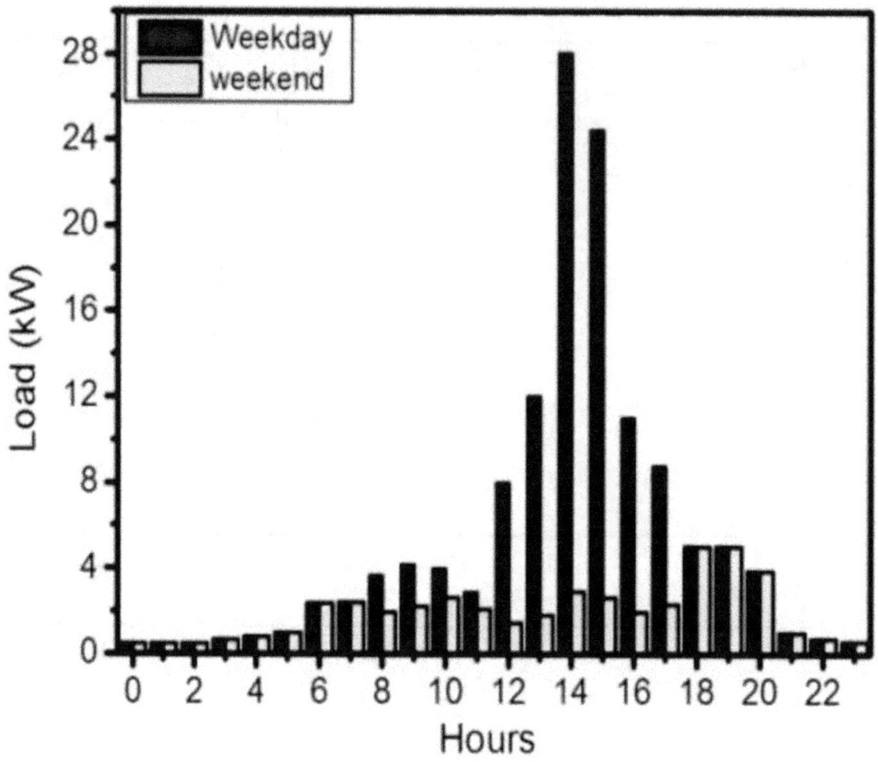

Comparison Between 0% UE and 9% UE Optimal Configurations

In this section, two optimal cases are compared together with respect to their COE, UE, EE, and the performances of the storage systems. Two cases are chosen first-0% UE (i.e.100% reliability) and second - least unmet energy and least excess energy (i.e. 9% UE 16.2% EE, as shown in Table 4 (ninth raw).

Figures 11 and 12 illustrate the hourly energy balanced (energy sharing) curve for 0% and 9% UE respectively for a sample week. In those seven days hourly simulations, the weekday and weekend loads are considered, hence it would be easy to analysis the unmet and dump load, and variations of solar radiation also. It can be easily seen that the optimal system for the 0% UE (Figure 11) has produced lots of excess energy (39.07%), because in this case the UR is not so enough to store that huge surplus energy produced

Figure 7. Daily average load demand variation for summer (weekday, weekend and overall)

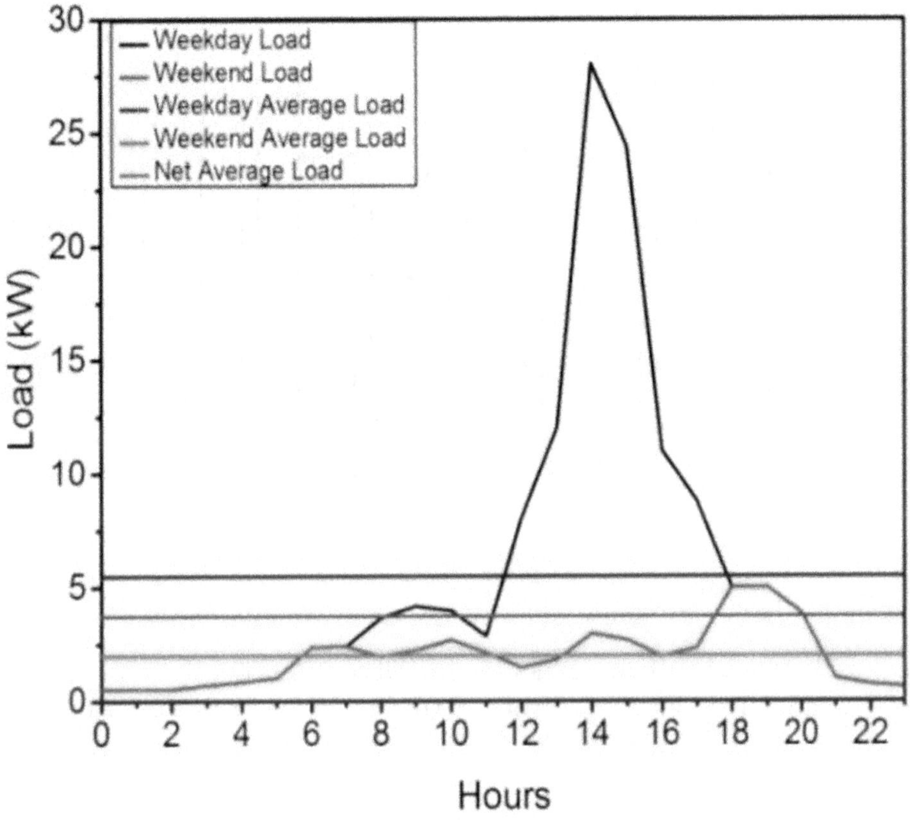

Figure 8. Monthly electricity demand variation

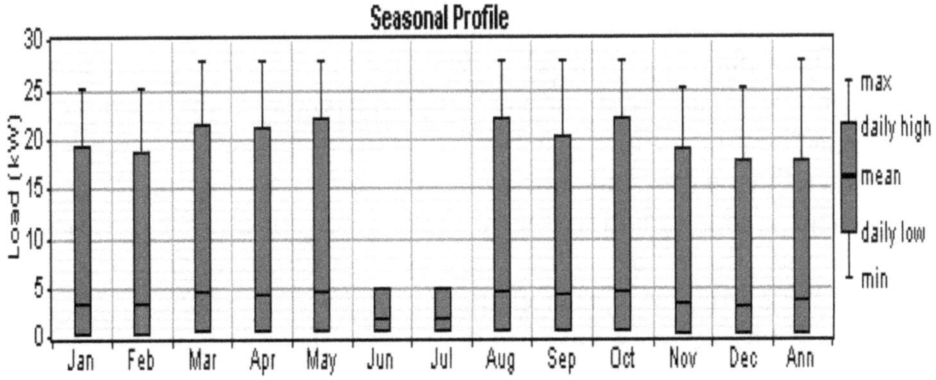

Figure 9. Average monthly GHI

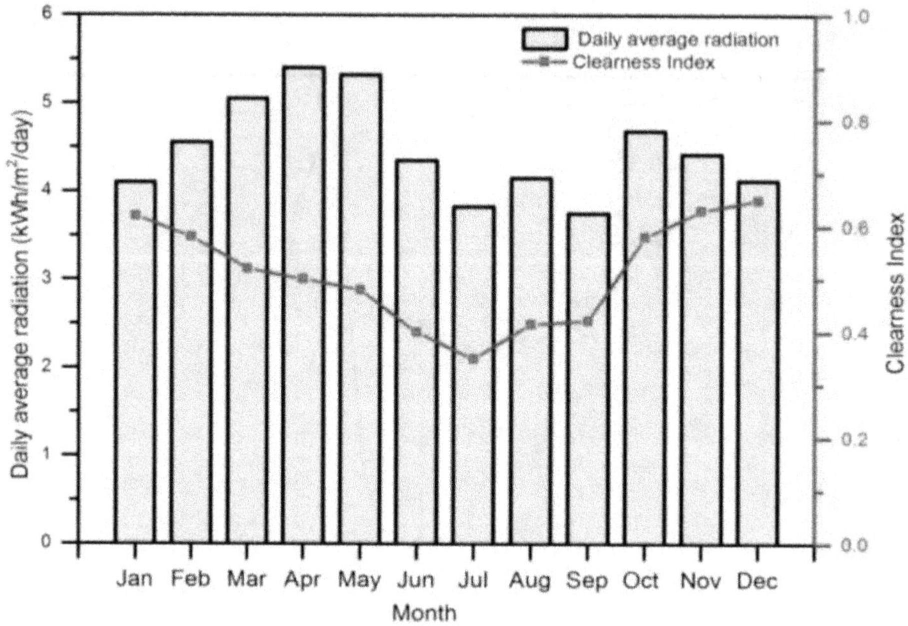

from the PV array. However, under the 9% UE optimal configuration, the % EE is dramatically reduced, i.e. from 39.07% to 16.2% (Table 4). Moreover, the volume of UR and size of PV plant are also greatly reduced, and hence COE has significantly declined. From Figure 11 (i.e. for 0% UE) it can be noted that there is a huge gap between load demand and PV power output, whereas for the 9% UE (Figure 12), the PV power and load power are almost at the same level. This is another reason why 9% UE can be considered as a benchmark level of unmet energy. Furthermore from Figure 11 and Figure 12, it can be observed that the excess energy is generally produced during 9:00 to 13:00 hours, because during this period, the load demand is generally low and solar radiation is at its peak. Also, it can be seen that on the 6th day of simulation, the solar irradiation is quite large (sunny day) as compared to load demand, and hence excess energy is produced in both the cases i.e. for 0% and 9% UE. However, the unmet energy (i.e. 9% UE "Figure 12") usually occurs when the solar radiation is very weak for two consecutive days. Figure 12 demonstrates that 3rd day of the week exhibits very poor solar insolation and hence UE energy has occurred after 13:00 Hrs.

Figure 13 and Figure 14 illustrate the UR's SOC probability distribution (relative frequency %) for 0% UE and 4% UE respectively. The results indicate

Figure 10. The hourly solar radiation variation

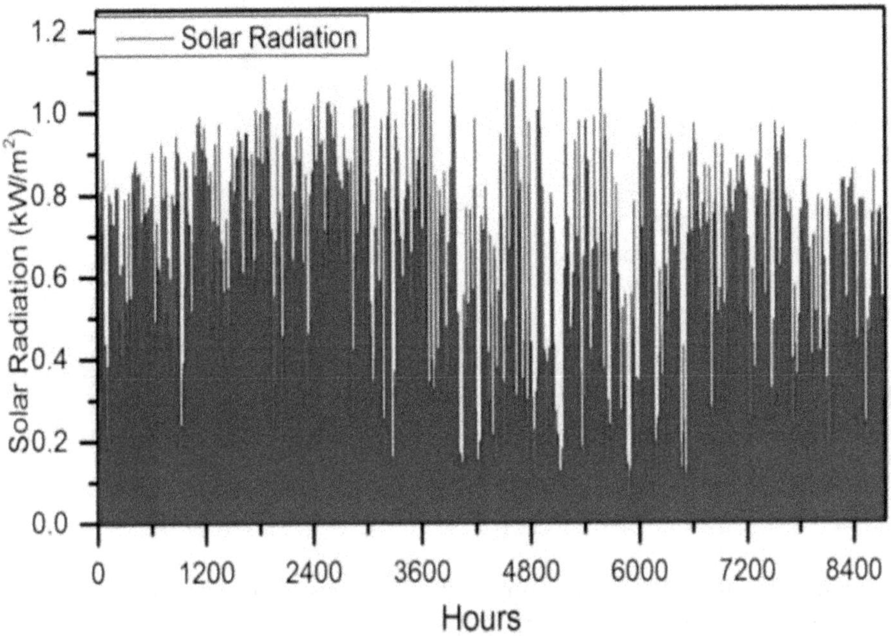

Table 9. System 1 configuration for the different UE (%) level

Trail No.	Input parameters				Output parameters	
	UE (%)	Size of PV-panel		Volume of UR (m^3)	EE (%)	COE ($/kWh)
		(kW)	Area (m^2)			
1	0	1.36	164.94	1153.43	2254.30	39.08
2	0.5	0.92	136.49	681.75	1370.36	37.75
3	1	0.85	124.81	872.78	1231.76	36.67
4	2	0.79	123.06	860.57	1065.25	36.92
5	3	0.72	79.90	558.72	1282.30	28.94
6	4	0.76	56.70	396.49	1678.49	19.09
7	5	0.69	55.15	385.65	1476.39	19.24
8	6	0.62	58.91	411.94	1185.63	22.26
9	7	0.62	79.91	558.78	937.50	30.62
10	8	0.60	49.41	345.54	1228.47	17.56
11	9	0.59	46.93	328.17	1188.29	16.19
12	10	0.56	66.97	468.31	882.86	28.08

Figure 11. Hourly energy sharing curve for a sample of week for 0% UE

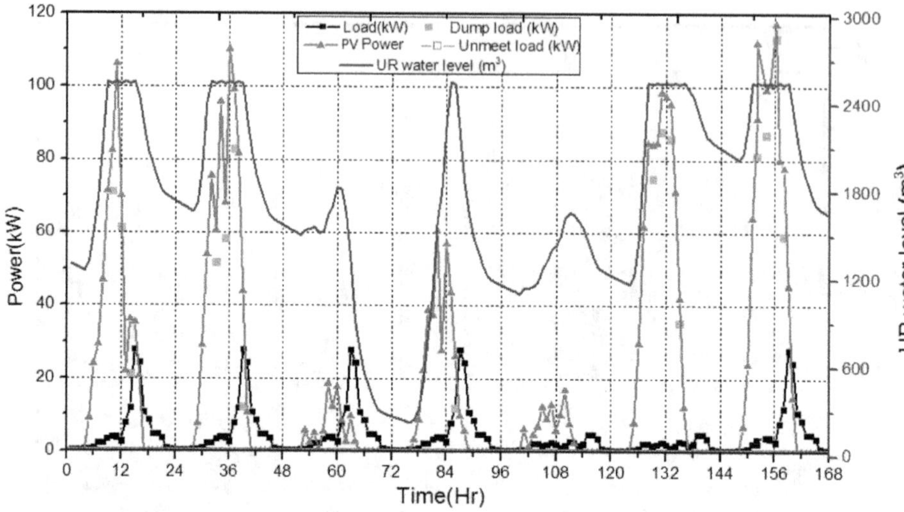

that about 90% of its SOC is equal to 40.17% frequency level under the 0% UE case. The probability of UR's SOC above 90% for a real PHS based PV plant project (Kaldellis, et al. 2019) is less than that of present PHS based PV system (32.7%). For 9% UE, high percentage relative frequency (i.e. 20.44%) lies between 90-100% of SOC, which is however less than that of real PHS

Figure 12. Hourly energy sharing curve for a sample of week for 9%UE

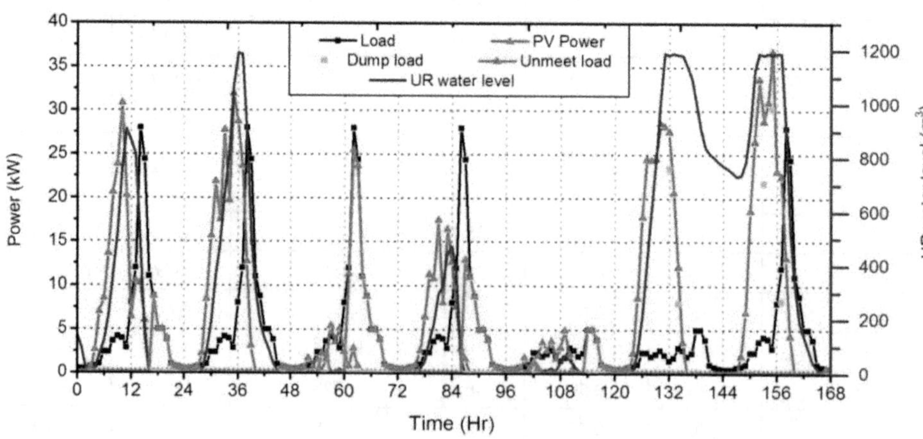

Figure 13. The relative frequency vs. upper reservoir's SOC (0%UE)

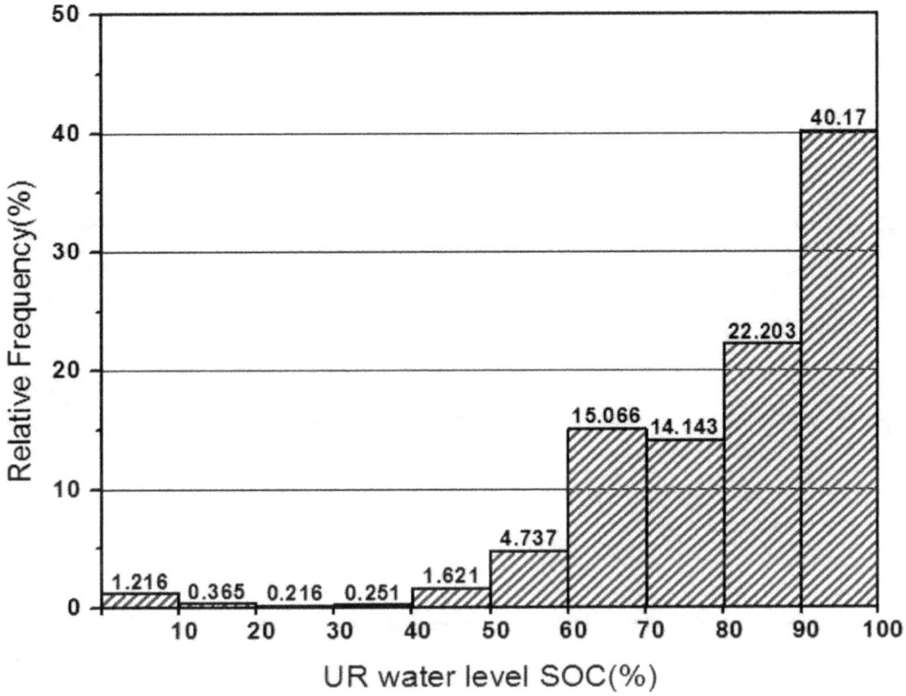

based PV system (Kaldellis, et al. 2019), and hence, performance of storage system has greatly improved.

Under the 100% power reliability optimal system (i.e. 0% UE), the total LCC is $770240, and for the 91% of reliability optimal system (i.e. 9% UE), the total LCC is $ 330050, which is a reduction of 57.15%. The system's LCC break-up by components is illustrated in Figure 15. It can be seen that, UR accounts for more than 60% of the system's total cost followed by the PV array (20.8%) for the zero unmet load. Although, under 9% UE, UR lead the system's cost, it is lower than that of the previous case (decreased by about 7.9%), and the percentage of PV cost increase by 7.64% as compared to 0% UE case.

Based on energy and water flow model, the daily mean energy and water flow for the optimal configuration (i.e. for the 9% UE optimal system) is summarized in Figure 16. In this case the total energy output of the PV array is a 61360.15 kWh/year and their peak power production is 42.10 kW, in which 25.56% (37.30 kWh/day) of PV's total energy directly satisfies the load demand, the rest of the demand is fulfilled by turbine (42.54 kWh/day), and

Figure 14. The relative frequency vs. upper reservoir's SOC (9%UE)

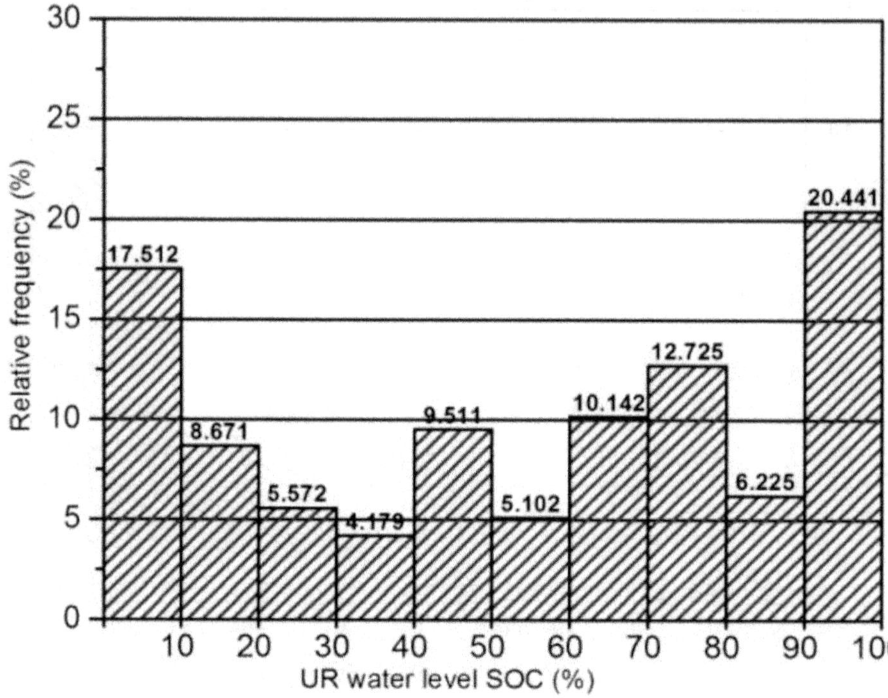

58.42% (85.24 kWh/day) of PV production goes to pump units. And 16.2% (23.85 kWh/day) of total PV output is dispatched to dump load, usually during the 11:00 am to 1:00 pm, and very small shortfall of energy occurs, which is

Figure 15. Break-down of LCC for 0% (left) and 9% (right) system1

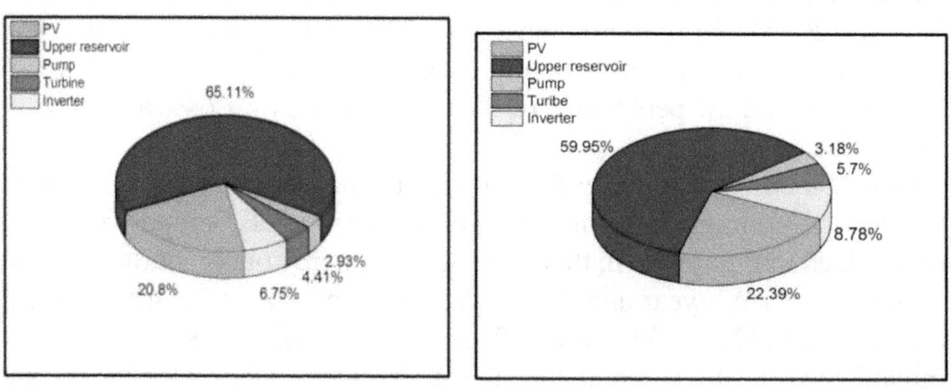

only 9% (7.96 kWh/day) that means 2905.4 kWh for the entire 8760 hours. The daily water flow rate elevated by pump to upper reservoir is 834.23m^3/day, whereas 743.31m^3/day of water is dispatched to the turbine, and some water leakage through evaporation also occurs; therefore daily mean round trip efficiency of PHS is about 50%.

Figure 16. Daily average electrical energy and water flow in the 4% UE optimal case

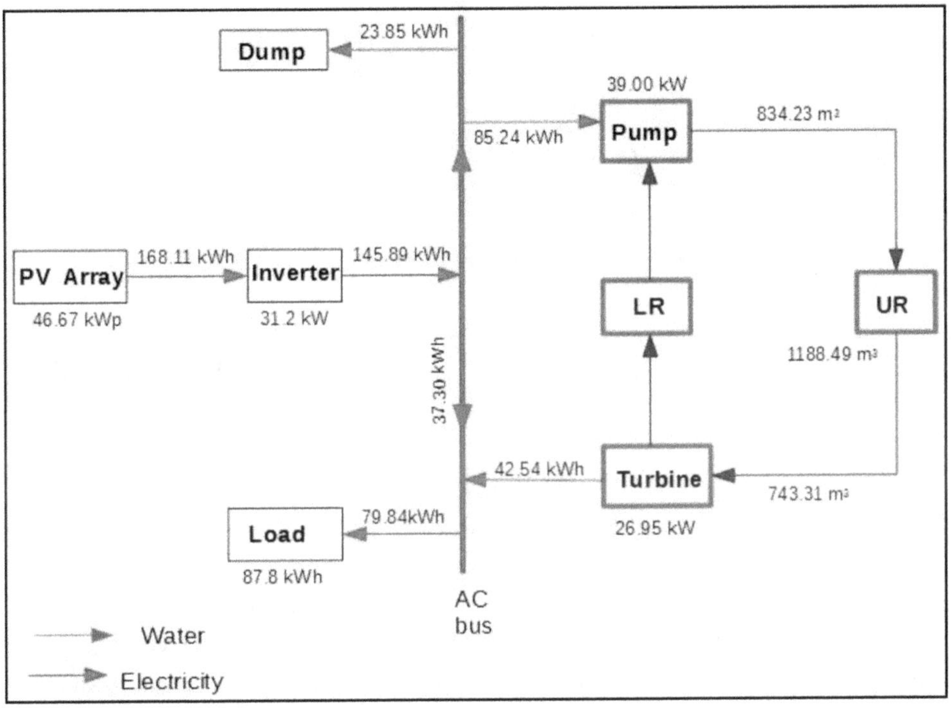

Comparison of SES's Load Profile With Existing Building's Load Profile

The SES' load profile generated by optimal solution for this academic building is illustrated in Figure 17. As can be seen from Figure 17, the surplus energy (during the sunny hours) goes to the storage unit, and this stored energy is dispatched during the night time. The load factor of the new load profile is 0.389, whereas the existing building's load factor is 0.131 that means the load factor has improved by 0.258.

Figure 17. RES's load and Building's load curve 9% UE

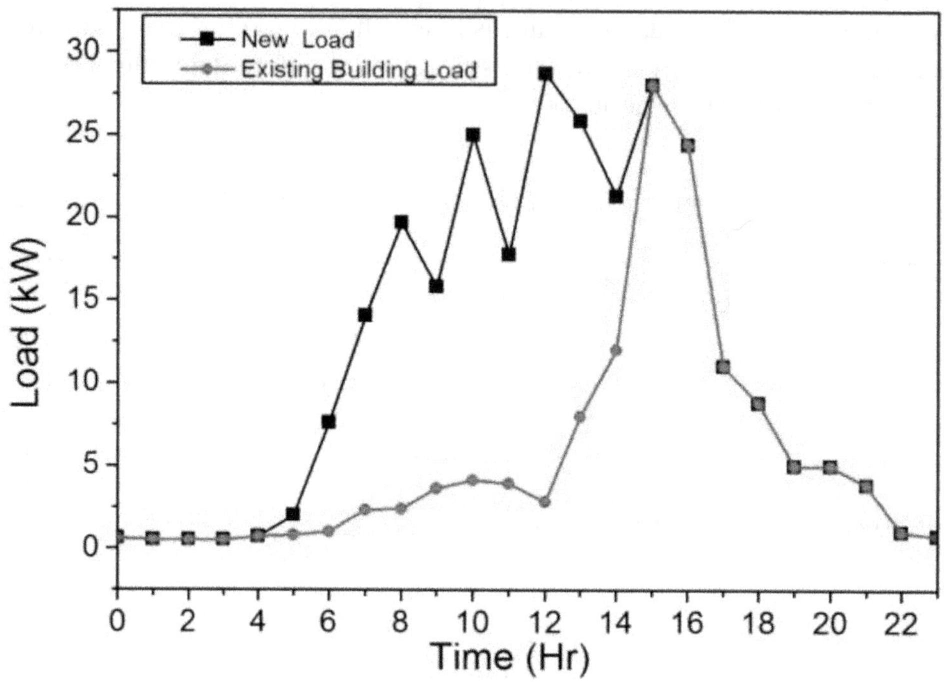

Optimization of SES Parameters

Finally, optimization of the SES system is performed based on the AV obtained via AHP-MOORA considering four independent and two dependent parameters with 12 nos. of trails. The optimization results are depicted in Table 10. Then, the ranking of the trail no is done based on the AV. The result shows that trail no.11 yields the highest rank with an AV of 0.7309 and the trial no.12 yields the least rank with relative index 0.3333, which is the worst trail result.

The optimal setting is obtained for Unmet Energy (UE) as 9%, Size of PV-panel as 46.93 kW and area 328.17 m^2, and volume of Upper Reservoir (UR) as 1188.29 m^3, which give optimal performance parameters i.e. Excess Energy (EE) of 16.19% and Cost of Energy (COE) of 0.59 \$/kWh. Thus, the recommended optimal setting provides optimal performances like higher efficiency and lower Excess Energy, which directly or indirectly increase the performance of the SES.

Table 10. Assessment values (AV) of SES system

Test No.	Assessment Values (*AV*)	Rank
1	0.4958	3
2	0.3690	10
3	0.4558	4
4	0.4134	7
5	0.4405	6
6	0.5073	2
7	0.4553	5
8	0.3907	8
9	0.3569	11
10	0.3845	9
11	0.7309	1
12	0.3333	12

Analysis of Variance Analysis

Further, Analysis of Variance (ANOVA) analysis with 95% of confidence interval is used to determine the significance and adequacy of the developed model. In the ANOVA analysis, the Fisher's statistical test (F-Ratio) and significance probability (P-value) followed by coefficient of determination (R^2) and adjusted-R^2 are determined to test the sufficiency and fitness of the proposed model. The F-value denotes the most influencing parameters on the performance characteristics while P-value shows the statistical significance of process variables of SES system. During the analysis, the process variables which have higher F-values and lower P-values ($P<0.05$) indicate statistical significance (Kulekci, et al., 2011). The R^2 value signifies the total variation in the response expected by the model. The purpose of adjusted-R^2 is to analyze the model adequacy and fitness. In this research stepwise elimination method is used to remove the insignificant parameters to adjust the fitted quadratic model (Rao, and Kalyankar, 2014). ANOVA analysis results are given in Table 11.

The ANOVA for AV value results show that the parameter UE and Volume of UR are the most significant factors influencing the SES performance. Size of PV-panel is considered as statistically insignificant as it fails the statistical significance test. Subsequently, the P values for UE and Volume of UR are

Table 11. ANOVA for AV values of SES system

Source	DF	Adj SS	Adj MS	F-Value	P-Value
Regression	4	0.027711	0.006928	0.51	0.009
Unmet Energy (UE)	1	0.000001	0.000001	0	0.003
Size of PV-panel area	1	0.018147	0.018147	1.35	0.823
Size of PV-panel volume	1	0.010029	0.010029	0.74	0.417
Upper Reservoir (UR)	1	0.000726	0.000726	0.05	0.003
Error	7	0.094351	0.013479		
Total	11	0.122062			
R^2=95.47%, R^2(adj)=96.78%, R^2 (Pred)= 97.58%					

found to be less than 0.05, which indicate that the model and associated terms are statistical significant (Konig, et al. 1985). The R^2 value is calculated as 0.9547 for AV values, which implies that 95% of experimental data are well-suited. Higher R^2 coefficient gives the satisfactory model with respect to the experimental data. Finally, the adequacy and fitness of the model is calculated by adjusted-R^2 values. The high value of adjusted-R^2 (0.9758 for AV values) supports a high correlation between the experimental and the predicted values (Kumar, et al., 2014).

Confirmatory Test

Optimal setting results obtained using the proposed method is verified through confirmatory tests. The optimal setting i.e. T_i= 34.1°C, I=535.60 W/m^2 at a constant m = 0.0031 kg/s; and A = 0.319 m^2, is used for confirmatory experiments and the corresponding results are shown in Table 12. The results show that confirmatory tests results are comparable and acceptable with experimental results for the optimal conditions.

Table 12. Confirmatory tests for SFPC thermal system

Input Conditions	Output Parameters	Experimental results	Confirmatory results
Test No-12: UE as 9%, Size of PV-panel as 46.93 kW and area 328.17 m^2, and volume of UR as 1188.29 m^3	EE in %	16.19	15.91
	COE in ($/ kWh	0.59	0.61

SUMMARY

The present study demonstrates the importance of multi-criterias like unmet energy, excess energy, components sizes, storage unit capacity, etc. in the technical and economic aspects (i.e. Cost of Energy) of the Solar Energy System. The present study also delineates the importance of optimal selection of type and storage unit capacity in the quest of bettering the SES design. An integrated MCDM method i.e. AHP-MOORA is proposed for modeling and optimization of sizing of the system. Here, the AHP method is used for extraction of priority weights of system parameters while MOORA for size optimization of SES parameters. Trails (12 nos.) have been performed on the Solar Energy System under different operating/input conditions. The optimization result indicates that trail no.11 produces the highest assessment values compared to the other trails. The optimal setting is obtained for Unmet Energy (UE) as 9%, Size of PV-panel as 46.93 kW and area 328.17 m^2, and volume of Upper Reservoir (UR) as 1188.29 m^3, which give optimal performance parameters i.e. Excess Energy (EE) of 16.19% and Cost of Energy (COE) of 0.59 $/kWh. The recommended optimal setting provides optimal performance like higher efficiency and lower Excess Energy which directly or indirectly increase the performance of the system.

Additionally, ANOVA analysis is carried out to determine the significance and adequacy of the developed model. The results show that the parameter Unmet Energy and Volume of Upper Reservoir are the major parameters and also model fitted data are well fitted with experimental result within 95% confidence level. At last, confirmatory tests show comparable and acceptable results with experimental values. Further, co-generation i.e. electrical energy and thermal energy may be made possible by using the optimized SES as the heat absorbed by the PV panel due to solar radiation absorbed by it can work as a source of heat that can be used by a coolant in a secondary circuit to generate solar thermal energy.

Finally, as the proposed MCDM technique gives optimized values of the considered parameters for the performance of the system, and as these are obtained with good accuracy and with statistical significance, the obtained optimized values can be practically implemented to fulfil the power consumption demand of an academic building. Hence, it is concluded that the performance of the system under the considered environment is found to be feasible and the proposed hybrid MCDM method provides acceptable results for the Solar Energy System.

REFERENCES

Belenov, A. T., Daus, Y. V., Rakitov, S. A., Yudaev, I. V., & Kharchenko, V. V. (2016). The experience of operation of the solar power plant on the roof of the administrative building in the town of Kamyshin, Volgograd oblast. *Applied Solar Energy*, *52*(2), 105–108. doi:10.3103/S0003701X16020092

Brauers, W. K. M., & Zavadskas, E. K. (2006). The MOORA method and its application to privatization in a transition economy. *Control and Cybernetics*, *35*(2), 445–469.

Chakraborty, S. (2011). Applications of the MOORA method for decision making in a manufacturing environment. *International Journal of Advanced Manufacturing Technology*, *54*(9-12), 1155–1166. doi:10.100700170-010-2972-0

George. (2002). Feasibility study of a hybrid wind/hydro power-system for low-cost electricity production. *Appl Energy*, *72*, 599–608.

Hoppmann, J., Volland, J., Schmidt, T. S., & Hoffmann, V. H. (2014). The economic viability of battery storage for residential solar photovoltaic systems–A review and a simulation model. *Renewable & Sustainable Energy Reviews*, *39*, 1101–1118. doi:10.1016/j.rser.2014.07.068

Jamil, M., Kirmani, S., & Rizwan, M. (2012). Techno-Economic Feasibility Analysis of Solar Photovoltaic Power Generation : A Review. *Smart Grid and Renewable Energy*, *03*(04), 266–274. doi:10.4236gre.2012.34037

Kaldellis, J. K., Kapsali, M., Kondili, E., & Zafirakis, D. (2013). Design of an integrated PV-based pumped hydro and battery storage system including desalination aspects for the Island of Tilos. *International Conference on clean electrical power, Alger, Sardinia, Italy.*

Karande, P., & Chakraborty, S. (2012). Application of multi-objective optimization on the basis of ratio analysis (MOORA) method for materials selection. *Materials & Design*, *37*(1), 317–324. doi:10.1016/j.matdes.2012.01.013

Konig W., Wulf C. & Willerscheid H. (1985). Machining of fibre reinforced plastics. *CIRP Annals Manufacturing Technology*, *34*, 537-548.

Kulekci, M. K., Esme, O. E., & Kazancoglu, Y. (2011). Modeling and prediction of weld shear strength in friction stir spot welding using design of experiments and neural network. *Materialwissenschaft und Werkstofftechnik*, *42*(11), 990–995. doi:10.1002/mawe.201100781

Kumar, R., Kumar, K., & Bhowmik, S. (2014). Optimization of mechanical properties of epoxy based wood dust reinforced green composite using taguchi method. *International Conference on Advances in Manufacturing and Materials Engineering*, *5*, 688-696. 10.1016/j.mspro.2014.07.316

Li, T., Yang, M., Zhao, L., & Wang, F. (2012). The research of wind-light complementary based on pumped storage power system. *Advanced Materials Research*, *354–355*, 1132–1136.

Ma, T., Yang, H., Lu, L., & Peng, J. (2014). Optimal design of an autonomous solar–wind-pumped storage power supply system. *Applied Energy*. Advance online publication. doi:10.1016/j.apenergy.2014.11.026

Ma, T., Yang, H., Lu, L., & Peng, J. (2014). Technical feasibility study on a standalone hybrid solar-wind system with pumped hydro storage for a remote island in Hong Kong. *Renewable Energy*, *69*, 7–15. doi:10.1016/j.renene.2014.03.028

Ma, T., Yang, H., Lu, L., & Peng, J. (2015). Pumped storage-based standalone photovoltaic power generation system: Modeling and techno-economic optimization. *Applied Energy*, *137*, 649–659. doi:10.1016/j.apenergy.2014.06.005

Mahmoud, M. M. (2014). On the storage batteries used in solar electric power systems and development of an algorithm for determining their ampere–hour capacity. *Electric Power Systems Research*, *71*(1), 85–89. doi:10.1016/j.epsr.2003.12.018

Manolakos, D., Papadakis, G., Papantonis, D., & Kyritsis, S. (2004). A stand-alone photovoltaic power system for remote villages using pumped water energy storage. *Energy*, *29*(1), 57–69. doi:10.1016/j.energy.2003.08.008

Nair, N. C., & Garimella, N. (2010). Battery energy storage systems: Assessment for small-scale renewable energy integration. *Energy and Building*, *42*(11), 2124–2130. doi:10.1016/j.enbuild.2010.07.002

Nesterenkov, P., & Kharchenko, V. (2018). Thermo Physical Principles of Cogeneration Technology with Concentration of Solar Radiation. *International Conference on Intelligent Computing and Optimization*, 117-128.

Panchenko, V., & Kharchenko, V. (2020). Development and Research of PVT Modules in Computer-Aided Design and Finite Element Analysis Systems. Handbook of Research on Energy-Saving Technologies for Environmentally-Friendly Agricultural Development, 314-342, doi:10.4018/978-1-5225-9420-8.ch013

Panchenko, V., Kharchenko, V., & Vasant, P. (2018). Modeling of solar photovoltaic thermal modules. *International Conference on Intelligent Computing and Optimization*, 108-116.

Rao, R. V., & Kalyankar, V. D. (2014). Optimization of modern machining processes using advanced optimization techniques-a review. *International Journal of Advanced Manufacturing Technology*, *73*(5-8), 1159–1188. doi:10.100700170-014-5894-4

R. Rawat, S.S. Chandel. (n.d.). Simulation and optimization of solar photovoltaic- wind stand-alone hybrid system in hilly terrain of India. *Int. J. of Renewable Energy Research*, *3*(3), 45-65.

Saaty, T. L. (1989). Decision making with the analytic hierarchy process. *Journal de Chirurgie*, *1*(1), 83–98.

Saini, R. P. (2011). *Large-scale energy storage systems*. IIT Roorkee.

Sarkar, A., Panja, S. C., Das, D., & Sarkar, B. (2015). Developing an efficient decision support system for non-traditional machine selection: An application of MOORA and MOOSRA. *Journal of Production and Manufacturing Research*, *3*(1), 324–342. doi:10.1080/21693277.2014.895688

Shaahid, S. M., & Elhadidy, M. A. (2007). Technical and economic assessment of grid independent hybrid photovoltaic-diesel-battery power system for commercial load in desert environments. *Renewable & Sustainable Energy Reviews*, *11*(8), 1794–1810. doi:10.1016/j.rser.2006.03.001

Singaravel, B., Selvaraj, T., & Vinod, S. (2016). Multi-objective optimization of turning parameters using the combined MOORA and Entropy method. *Transactions of the Canadian Society for Mechanical Engineering*, *40*(1), 101–111. doi:10.1139/tcsme-2016-0008

Tzeng, G. H., & Huang, J. J. (2011). *Multiple attribute decision making: methods and applications*. CRC Press. doi:10.1201/b11032

Wohlgemuth, J. H. (2012). Standards for PV Modules and Components– Recent Developments and Challenges. *27th European Photovoltaic Solar Energy Conference and Exhibition Frankfurt, Germany,* NREL/CP-5200-56531.

Appendix

List of Nomenclature

f_{max} Largest eigenvalue of fuzzy decision matrix
g Number of criterias to be maximized
m Number of alternatives
n number of criterias
n-g Number of criterias to be minimized
w Weighted normalized decision matrix of ith alternatives on jth criterion for beneficial criteria
w* Weighted normalized decision matrix of ith alternatives on jth criterion for non-beneficial criteria
F_R Collector heat removal factor
CI_f Fuzzy consistency index
CR_f Fuzzy consistency ratio
N_{ij} Normalized matrix of ith alternatives on jth criterion for beneficial criteria
N_{ij}* Normalized matrix of ith alternatives on jth criterion for non-beneficial criteria
RPI_i Relative performance index of ith alternative with respect to all criterias
ξ_h Weights of higher the better parameters for the ith parameter to corresponding jth parameter
$\xi_{l\,W}$eights of lower the better parameters for the ith parameter to corresponding jth parameter
Ψ_1 weighted decision matrix for higher the better parameters
Ψ_2 weighted decision matrix for lower the better parameters
T_{adm} Average ambient temperature during the mth time interval (°C)
T_{sfm}/T_{sim} Water temperature recorded by RTDs at the end and start of the mth time interval (°C)
G_{Tfm}/G_{Tim} Solar irradiance value at the end and at the start of the mth time interval (W/m^2)

A_{ij} Decision matrix for *j*th criteria to *i*th alternative
G_{ij} / K_{ij} Normalization matrix of *i*th alternatives on *j*th criterion
R_j Relative index value of *j*th criterion
D_{ij} Decision matrix of performance value of *j*th criteria to *i*th alternative
d_j Degree of divergence for each of the *j*th criteria
D_{bigger} Beneficial element of the column
$D_{smaller}$ Non-beneficial element of the column
R_j Entropy matrix of *i*th alternatives on *j*th criterion
ζ_j Precise priority weights for each of *j*th criterion
W_{ij} Weighted normalized decision matrix of *i*th alternatives on *j*th criterion

List of Greek Symbol

T_a Ambient temperature in °C
T_i Inlet water temperature in °C
T_o Outlet water temperature in °C
U_L Collector heat loss coefficient in W/m^2 °C
W_s Wind velocity in m/s
c Height of the mirror in m
h_s Height of receiver in summer, in m
h_w Height of receiver in winter, in m
t Time in min
G Global solar radiation, in W/m^2
L Distance of receiver from a heliostat mirror in m
T Temperature, in °C
T_{sfd} Final storage water temperature in °C
A_c Area of the collector in m^2
T_{sid} Initial value of storage water temperature in °C
H Useful heat gain, in W
$\dot{m}$ Mass flow rate, in kg/s
A Absorber area, in m^2
I Solar irradiance, in W/m^2
β Heliostat tilt angle in deg.
θ Incident angle in deg.
α Altitude angle in deg.
η_{th} Thermal efficiency in %
G_T Global solar radiation on tilted surface in W/m^2

W_s Wind speed in m/s
η_{sys} System efficiency in %

List of Abbreviation

AHP Analytic Hierarchy Process
ANOVA Analysis of Variance
AV Assessment Value
CPC Compound Parabolic Collector
CRS Central Receiver System
CST Concentrated Solar Thermal
DEA Data Envelopment Analysis
DEAR Data Envelopment Analysis based Ranking
DM Decision Maker
ETSTC Evacuated Tubular Solar Thermal Collector
FPC Flat Plate Collector
GRA Grey Relational Analysis
GTMA Graph Theory and Matrix Approach
HFC Heliostat Field Collector
LFR Linear Fresnel Reflector
MADM Multi-Attribute Decision Making
MCDM Multi-Criteria Decision Making
MODM Multi-Objective Decision Making
MOORA Multi-Objective Optimization on the bases of Ratio Analysis
MOOSRA Multi-Objective Optimization based on Simple Ratio Analysis
MPRI Multiple Performance Ranking Index
PCM Phase Change Materials
PDR Parabolic Dish Reflector
PHS Pumped Hydro Energy Storage
PSI Process Selection Index
PTC Parabolic Trough Collector
PV Photovoltaic
RI Random Index
RPI Relative Performance Index
RTD Resistance Temperature Detector
SAW Simple Additive Weighting
SFPC Solar Flat Plate Collector

TFN Triangular Fuzzy Number
TOPSIS Technique for Order of Preference by Similarity to Ideal Solution

Related Readings

To continue IGI Global's long-standing tradition of advancing innovation through emerging research, please find below a compiled list of recommended IGI Global book chapters and journal articles in the areas of solar thermal systems, parabolic trough collectors, and hybrid solar energy. These related readings will provide additional information and guidance to further enrich your knowledge and assist you with your own research.

Abbadi, A., Hamidia, F., Morsli, A., Bellatrache, H., Boukhetala, D., & Nazli, L. (2020). Interval Type 2 Fuzzy-Logic-Based Solar Power MPPT Algorithm Connected to AC Grid. *International Journal of Energy Optimization and Engineering*, *9*(1), 110–121. doi:10.4018/IJEOE.2020010103

Abdullah, L., & Najib, L. (2017). Interval Type-2 Fuzzy Analytic Hierarchy Process for Sustainable Energy Sources Selection. *International Journal of Fuzzy System Applications*, *6*(3), 124–137. doi:10.4018/IJFSA.2017070106

Affam, A. C., & Chaudhuri, M. (2019). Comparative Study of Advance Oxidation Processes for Treatment of Pesticide Wastewater. In H. A. Aziz & S. S. Abu Amr (Eds.), *Advanced Oxidation Processes (AOPs) in Water and Wastewater Treatment* (pp. 261–323). IGI Global. doi:10.4018/978-1-5225-5766-1.ch011

Akkucuk, U. (2018). Consumer Attitudes Towards Renewable Energy. In U. Akkucuk (Ed.), *Handbook of Research on Supply Chain Management for Sustainable Development* (pp. 61–74). IGI Global. doi:10.4018/978-1-5225-5757-9.ch004

Ali, A. Nor, N. M. Ibrahim, T. Romlie, M. F. & Bingi, K. (2018). Sizing and Placement of Battery-Sourced Solar Photovoltaic (B-SSPV) Plants in Distribution Networks. In Singh, U. P. Tiwari, A. & Singh, R. K. (Ed.), Soft-Computing-Based Nonlinear Control Systems Design (pp. 220-251). IGI Global. http://doi:10.4018/978-1-5225-3531-7.ch011

Ali, A., Nor, N. M., Ibrahim, T., Romlie, M. F., & Bingi, K. (2018). Big Data Storage for the Modeling of Historical Time Series Solar Irradiations. In R. S. Segall & J. S. Cook (Eds.), *Handbook of Research on Big Data Storage and Visualization Techniques* (pp. 433–463). IGI Global. doi:10.4018/978-1-5225-3142-5.ch016

Ali, D. A., Palaniandy, P., & Feroz, S. (2019). Advanced Oxidation Processes (AOPs) to Treat the Petroleum Wastewater. In H. A. Aziz & S. S. Abu Amr (Eds.), *Advanced Oxidation Processes (AOPs) in Water and Wastewater Treatment* (pp. 99–122). IGI Global. doi:10.4018/978-1-5225-5766-1.ch005

Aslani, A., Niknejad, M., & Maghami, A. (2017). Robustness of US Economy and Energy Supply/Demand Fluctuations. *International Journal of Energy Optimization and Engineering*, *6*(4), 1–15. doi:10.4018/IJEOE.2017100101

Aziz, N. A., & Palaniandy, P. (2019). Photocatalysis (TiO2/Solar) in Water and Wastewater Treatment. In H. A. Aziz & S. S. Abu Amr (Eds.), *Advanced Oxidation Processes (AOPs) in Water and Wastewater Treatment* (pp. 171–199). IGI Global. doi:10.4018/978-1-5225-5766-1.ch008

Babaev, B. D., Kharchenko, V., Panchenko, V., & Vasant, P. (2019). Materials and Methods of Thermal Energy Storage in Power Supply Systems. In V. Kharchenko & P. Vasant (Eds.), *Renewable Energy and Power Supply Challenges for Rural Regions* (pp. 115–135). IGI Global. doi:10.4018/978-1-5225-9179-5.ch005

Babaev, B. D., Kharchenko, V. V., & Panchenko, V. (2020). Development and Research of Phase-Transition and Thermochemical Materials for Heat Accumulation. In V. Kharchenko & P. Vasant (Eds.), *Handbook of Research on Smart Computing for Renewable Energy and Agro-Engineering* (pp. 1–26). IGI Global. doi:10.4018/978-1-7998-1216-6.ch001

Babatunde, D. E., Denwigwe, I. H., Babatunde, O. M., Agboola, O., & Akinsipe, G. D. (2019). Relevance of Chemically Functionalized Nano-Fillers and Modified Nanocomposite in Energy Systems. In N. Ramdani (Ed.), *Polymer Nanocomposites for Advanced Engineering and Military Applications* (pp. 10–65). IGI Global. doi:10.4018/978-1-5225-7838-3.ch002

Badis, A., Boujmil, M. H., & Mansouri, M. N. (2018). A Comparative Study on Maximum Power Point Tracking Techniques of Photovoltaic Systems. *International Journal of Energy Optimization and Engineering*, *7*(1), 66–85. doi:10.4018/IJEOE.2018010104

Badruddin, S., Robertson-Gillis, C. R., Ashworth, J., & Wright, D. J. (2020). Economically Optimal Solar Power Generation. In P. Yang (Ed.), *Cases on Green Energy and Sustainable Development* (pp. 254–290). IGI Global. doi:10.4018/978-1-5225-8559-6.ch010

Bagdadee, A. H., & Zhang, L. (2019). A Review of the Smart Grid Concept for Electrical Power System. *International Journal of Energy Optimization and Engineering*, *8*(4), 105–126. doi:10.4018/IJEOE.2019100105

Balsas, C. J. (2019). Revaluing Saltscapes in Portugal. In D. Batabyal & D. K. Das (Eds.), *Global Trends, Practices, and Challenges in Contemporary Tourism and Hospitality Management* (pp. 31–57). IGI Global. doi:10.4018/978-1-5225-8494-0.ch002

Baporikar, N., & Fotolela, R. (2018). E-marketing for SMEs. *International Journal of Applied Management Sciences and Engineering*, *5*(1), 11–28. doi:10.4018/IJAMSE.2018010102

Barisal, A. K., & Lal, D. K. (2018). Application of Moth Flame Optimization Algorithm for AGC of Multi-Area Interconnected Power Systems. *International Journal of Energy Optimization and Engineering*, *7*(1), 22–49. doi:10.4018/IJEOE.2018010102

Belay, T. (2020). Micro-Grid Planning and Resilience Within Bulk System Planning and Operation. In B. Khan, H. H. Alhelou, & G. Hayek (Eds.), *Handbook of Research on New Solutions and Technologies in Electrical Distribution Networks* (pp. 232–247). IGI Global. doi:10.4018/978-1-7998-1230-2.ch013

Ben Smida, M., Sakly, A., Vaidyanathan, S., & Azar, A. T. (2018). Control-Based Maximum Power Point Tracking for a Grid-Connected Hybrid Renewable Energy System Optimized by Particle Swarm Optimization. In A. Azar & S. Vaidyanathan (Eds.), *Advances in System Dynamics and Control* (pp. 58–89). IGI Global. doi:10.4018/978-1-5225-4077-9.ch003

Bennamoun, L. (2019). Using Diffusion Model for Prediction and Optimization of Drying Process of Building Material. In M. Belgasmia (Ed.), *Optimization of Design for Better Structural Capacity* (pp. 1–23). IGI Global. doi:10.4018/978-1-5225-7059-2.ch001

Bergander, M. J., & Kapayeva, S. D. (2019). Solar Refrigeration for Post-Harvest Storage of Agricultural Products. In P. D. Gaspar & P. D. da Silva (Eds.), *Novel Technologies and Systems for Food Preservation* (pp. 108–139). IGI Global. doi:10.4018/978-1-5225-7894-9.ch005

Bhongade, S., & Verma, R. (2020). Optimized Hybrid Power System Using Superconducting Magnetic Energy Storage System. In S. Shandilya, S. K. Shandilya, T. Thakur, & A. K. Nagar (Eds.), *Novel Advancements in Electrical Power Planning and Performance* (pp. 28–77). IGI Global. doi:10.4018/978-1-5225-8551-0.ch002

Bhowmik, C., Dhar, S., & Ray, A. (2019). Comparative Analysis of MCDM Methods for the Evaluation of Optimum Green Energy Sources. *International Journal of Decision Support System Technology*, *11*(4), 1–28. doi:10.4018/IJDSST.2019100101

Boteanu, C. M. (2018). Improving Efficiency of the Romanian Energy Sector by Supporting Investments in Renewable Energy Use. *International Journal of Innovation in the Digital Economy*, *9*(3), 39–51. doi:10.4018/IJIDE.2018070104

Bottaccioli, L., Patti, E., Osello, A., Cerquitelli, T., Macii, E., & Acquaviva, A. (2019). IoT Platforms and Technologies Driving Spatial Planning and Analytics. In A. Voghera & L. La Riccia (Eds.), *Spatial Planning in the Big Data Revolution* (pp. 84–112). IGI Global. doi:10.4018/978-1-5225-7927-4.ch004

Butuzov, V. A., Butuzov, V. V., Bryantceva, E., & Gnatyuk, I. (2018). Experience and Prospects of Using Solar Energy for Heating Supply in Russia. In V. Kharchenko & P. Vasant (Eds.), *Handbook of Research on Renewable Energy and Electric Resources for Sustainable Rural Development* (pp. 26–60). IGI Global. doi:10.4018/978-1-5225-3867-7.ch002

Cali, U., & Lima, C. (2020). Energy Informatics Using the Distributed Ledger Technology and Advanced Data Analytics. In P. Yang (Ed.), *Cases on Green Energy and Sustainable Development* (pp. 438–481). IGI Global. doi:10.4018/978-1-5225-8559-6.ch016

Cebeci, U. DÃ1/4zgit, Z. & Ã‡avdar, C. (2020). An Innovative Solar Energy and Decision Support System Approach by Using IoT. In Ekren, G. Erkollar, A. & Oberer, B. (Ed.), Advanced MIS and Digital Transformation for Increased Creativity and Innovation in Business (pp. 169-187). IGI Global. http://doi:10.4018/978-1-5225-9550-2.ch008

Chaabna, A., Babouri, A., Huang, C., & Zhang, X. (2019). Visible Light Communication System for Indoor Positioning Using Solar Cell as Receiver. *International Journal of Energy Optimization and Engineering*, *8*(2), 47–60. doi:10.4018/IJEOE.2019040103

Chander, S., & Jangid, N. (2020). *Applications of PANI Thin Films*. IGI Global. doi:10.4018/978-1-5225-9896-1.ch006

Chander, S., & Jangid, N. (2020). *Conclusions and Future Aspects*. IGI Global. doi:10.4018/978-1-5225-9896-1.ch008

Chander, S., & Jangid, N. (2020). *PANI Thin Films for Solar Cells*. IGI Global. doi:10.4018/978-1-5225-9896-1.ch007

Chaudhry, S., & Garg, S. (2019). Smart Irrigation Techniques for Water Resource Management. In R. C. Poonia, X. Gao, L. Raja, S. Sharma, & S. Vyas (Eds.), *Smart Farming Technologies for Sustainable Agricultural Development* (pp. 196–219). IGI Global. doi:10.4018/978-1-5225-5909-2.ch009

Chawla, R., Singhal, P., & Garg, A. K. (2019). Impact of Dust for Solar PV in Indian Scenario. In K. Ahuja & A. Khosla (Eds.), *Driving the Development, Management, and Sustainability of Cognitive Cities* (pp. 111–138). IGI Global. doi:10.4018/978-1-5225-8085-0.ch005

Chawla, Y. P., & Singh, R. S. (2017). Skill Space Mission India 2025. In U. C. Okolie & A. M. Yasin (Eds.), *Technical Education and Vocational Training in Developing Nations* (pp. 214–235). IGI Global. doi:10.4018/978-1-5225-1811-2.ch010

Chiagozie, O. G., Okafor, K. C., & Nwaokolo, F. I. (2020). Robotic Expert System for Energy Management in Distributed Grid Ecosystem. *International Journal of Energy Optimization and Engineering*, *9*(1), 1–26. doi:10.4018/IJEOE.2020010101

Chirambo, D. (2020). Microfinance and Polycentric Governance as Strategies for Renewable Energy Deployment in Urban Sub-Saharan Africa. In P. Yang (Ed.), *Cases on Green Energy and Sustainable Development* (pp. 113–141). IGI Global. doi:10.4018/978-1-5225-8559-6.ch005

Das, D., Verma, R., & Pathak, V. (2019). Plasma Sprayed WC-12%Co-Coatings for TBC Applications on Diesel Engine Piston. *International Journal of Surface Engineering and Interdisciplinary Materials Science*, *7*(1), 37–54. doi:10.4018/IJSEIMS.2019010103

Daus, Y., Kharchenko, V., & Yudaev, I. V. (2018). Solar Radiation Intensity Data as Basis for Predicting Functioning Modes of Solar Power Plants. In V. Kharchenko & P. Vasant (Eds.), *Handbook of Research on Renewable Energy and Electric Resources for Sustainable Rural Development* (pp. 283–309). IGI Global. doi:10.4018/978-1-5225-3867-7.ch012

Daus, Y., Kharchenko, V., Yudaev, I. V., Dyachenko, V., & Klychev, S. (2019). Designing Solar Power Plant Due to Consumer Load Schedule, Solar Energy Potential, and Electricity Prime Cost. In V. Kharchenko & P. Vasant (Eds.), *Renewable Energy and Power Supply Challenges for Rural Regions* (pp. 168–196). IGI Global. doi:10.4018/978-1-5225-9179-5.ch008

Daus, Y., Kharchenko, V., Yudaev, I. V., Dyachenko, V., Klychev, S., & Rakitov, S. (2020). Potential Evaluation and Best Practices of Solar Power Plant Application in Rural Areas. In V. Kharchenko & P. Vasant (Eds.), *Handbook of Research on Energy-Saving Technologies for Environmentally-Friendly Agricultural Development* (pp. 293–313). IGI Global. doi:10.4018/978-1-5225-9420-8.ch012

Dempster, A. G. (2018). GNSS Vulnerability. In K. Yu (Ed.), *Positioning and Navigation in Complex Environments* (pp. 515–531). IGI Global. doi:10.4018/978-1-5225-3528-7.ch013

Deo, R. C. Ghimire, S. Downs, N. J. & Raj, N. (2018). Optimization of Windspeed Prediction Using an Artificial Neural Network Compared With a Genetic Programming Model. In Kim, D. Sekhar Roy, S. LÃ¤nsivaara, T. Deo, R. & Samui, P. (Ed.), Handbook of Research on Predictive Modeling and Optimization Methods in Science and Engineering (pp. 328-359). IGI Global. http://doi:10.4018/978-1-5225-4766-2.ch015

Deogam, N., & Shabbiruddin. (2019). Optimal Site Selection for Solar Photovoltaic Power Plant in North Eastern State of India using Hybrid MCDM Tools. *International Journal of Energy Optimization and Engineering*, *8*(2), 61–84. doi:10.4018/IJEOE.2019040104

Di Girolamo, G. (2018). Current Challenges and Future Perspectives in the Field of Thermal Barrier Coatings. In A. H. Pakseresht (Ed.), *Production, Properties, and Applications of High Temperature Coatings* (pp. 25–59). IGI Global. doi:10.4018/978-1-5225-4194-3.ch002

Dincer, H., Yuksel, S., & Canbolat, Z. N. (2019). A Strategic Approach to Reduce Energy Imports of E7 Countries. In R. C. Das (Ed.), *Handbook of Research on Economic and Political Implications of Green Trading and Energy Use* (pp. 18–38). IGI Global. doi:10.4018/978-1-5225-8547-3.ch002

Diolatzis, I. S., & Pavlogeorgatos, G. (2019). A Digital (3D) Reconstruction of a Lost Planetarium. *International Journal of Computational Methods in Heritage Science*, *3*(2), 91–117. doi:10.4018/IJCMHS.2019070106

Dzobo, O., & Awodele, K. O. (2020). Probabilistic Power System Reliability Assessment. In S. Shandilya, S. K. Shandilya, T. Thakur, & A. K. Nagar (Eds.), *Novel Advancements in Electrical Power Planning and Performance* (pp. 94–117). IGI Global. doi:10.4018/978-1-5225-8551-0.ch004

Elsayed, A. M., Dakkama, H. J., Mahmoud, S., Al-Dadah, R., & Kaialy, W. (2017). Sustainable Cooling Research Using Activated Carbon Adsorbents and Their Environmental Impact. In T. Kobayashi (Ed.), *Applied Environmental Materials Science for Sustainability* (pp. 186–221). IGI Global. doi:10.4018/978-1-5225-1971-3.ch009

Fendri, D., & Chaabene, M. (2020). Application of Hybrid Petri Nets for the Energy Dispatching of an Isolated Micro-Grid. *International Journal of Applied Metaheuristic Computing*, *11*(1), 61–72. doi:10.4018/IJAMC.2020010105

Finco, G. (2017). Energy in 3D. In G. Panconesi & M. Guida (Eds.), *Handbook of Research on Collaborative Teaching Practice in Virtual Learning Environments* (pp. 444–463). IGI Global. doi:10.4018/978-1-5225-2426-7.ch023

Fokides, E., Mastrokoukou, A., & Atsikpasi, P. (2017). Teaching Basic Astronomy Concepts to Pre-Service Teachers Using 3D Virtual Environments. In P. Tripathi & S. Mukerji (Eds.), *Handbook of Research on Technology-Centric Strategies for Higher Education Administration* (pp. 175–194). IGI Global. doi:10.4018/978-1-5225-2548-6.ch011

Ganesan, T. (2020). *Computational Intelligence in Energy Generation.* IGI Global. http://doi:10.4018/978-1-7998-1710-9.ch001

Ganesan, T., Vasant, P., & Elamvazuthi, I. (2017). Multiobjective Optimization of Solar-Powered Irrigation System with Fuzzy Type-2 Noise Modelling. In A. K. Adak, D. Manna, & M. Bhowmik (Eds.), *Emerging Research on Applied Fuzzy Sets and Intuitionistic Fuzzy Matrices* (pp. 189–214). IGI Global. doi:10.4018/978-1-5225-0914-1.ch008

Geleta, D. K., & Manshahia, M. S. (2020). Artificial Bee Colony-Based Optimization of Hybrid Wind and Solar Renewable Energy System. In V. Kharchenko & P. Vasant (Eds.), *Handbook of Research on Energy-Saving Technologies for Environmentally-Friendly Agricultural Development* (pp. 429–453). IGI Global. doi:10.4018/978-1-5225-9420-8.ch017

Ghose, D., Naskar, S., Shabbiruddin, & Roy, A. K. (2019). An Open Source Software. *International Journal of Open Source Software and Processes, 10*(1), 49–68. doi:10.4018/IJOSSP.2019010104

Gibberd, J. (2020). Green Building Technologies. In P. Yang (Ed.), *Cases on Green Energy and Sustainable Development* (pp. 482–510). IGI Global. doi:10.4018/978-1-5225-8559-6.ch017

Giostra, S. (2018). Energy/Form/Rule. In D. D'Uva (Ed.), *Handbook of Research on Form and Morphogenesis in Modern Architectural Contexts* (pp. 404–431). IGI Global. doi:10.4018/978-1-5225-3993-3.ch019

Gok, S. G. (2019). Turkey's Renewable Energy Potential. In A. S. Ikiz (Ed.), *Economic Dynamics of Global Energy Geopolitics* (pp. 83–95). IGI Global. doi:10.4018/978-1-5225-4203-2.ch005

Hassanzadeh, M., Saremi, M., Valefi, Z., & Pakseresht, A. H. (2018). Investigation on Improved-Durability Thermal Barrier Coatings. In A. H. Pakseresht (Ed.), *Production, Properties, and Applications of High Temperature Coatings* (pp. 60–78). IGI Global. doi:10.4018/978-1-5225-4194-3.ch003

Hazra, S., Pal, T., & Roy, P. K. (2019). Renewable Energy Based Economic Emission Load Dispatch Using Grasshopper Optimization Algorithm. *International Journal of Swarm Intelligence Research, 10*(1), 38–57. doi:10.4018/IJSIR.2019010103

Henao, J., & Sotelo, O. (2018). Surface Engineering at High Temperature. In A. H. Pakseresht (Ed.), *Production, Properties, and Applications of High Temperature Coatings* (pp. 131–159). IGI Global. doi:10.4018/978-1-5225-4194-3.ch006

Herrera, M. M., Carvajal-Prieto, L. A., Uriona-Maldonado, M., & Ojeda, F. (2019). Modeling the Customer Value Generation in the Industry's Supply Chain. *International Journal of System Dynamics Applications, 8*(4), 1–13. doi:10.4018/IJSDA.2019100101

Ilic, B. S. (2019). Renewable Energy Sources Management and Role of Ecological Parks. In U. Akkucuk (Ed.), *Ethical and Sustainable Supply Chain Management in a Global Context* (pp. 220–228). IGI Global. doi:10.4018/978-1-5225-8970-9.ch014

Iqbal, F., Rana, A. S., & Wani, S. A. (2018). Design and Analysis of a Cost-Effective Standalone Solar. In P. Kumar, S. Singh, I. Ali, & T. Ustun (Eds.), *Handbook of Research on Power and Energy System Optimization* (pp. 552–570). IGI Global. doi:10.4018/978-1-5225-3935-3.ch016

Jagatheesan, K., & Anand, B. (2018). AGC of Multi-Area Hydro-Thermal Power Systems With GRC Non-Linearity and Classical Controller. *Journal of Global Information Management, 26*(3), 11–24. doi:10.4018/JGIM.2018070102

Jagatheesan, K., Anand, B., Dey, N., & Ashour, A. S. (2018). Effect of SMES Unit in AGC of an Interconnected Multi-Area Thermal Power System With ACO-Tuned PID Controller. In N. Dey (Ed.), *Advancements in Applied Metaheuristic Computing* (pp. 164–184). IGI Global. doi:10.4018/978-1-5225-4151-6.ch007

Jain, R., Goel, V., Rekhi, J. K., & Alzubi, J. A. (2020). IoT-Based Green Building. In A. Solanki & A. Nayyar (Eds.), *Green Building Management and Smart Automation* (pp. 184–207). IGI Global. doi:10.4018/978-1-5225-9754-4.ch009

Jain, R., & Nagia, K. (2019). Smart Roads and Parking. In D. Goyal, S. Balamurugan, S. Peng, & D. S. Jat (Eds.), *The IoT and the Next Revolutions Automating the World* (pp. 168–185). IGI Global. doi:10.4018/978-1-5225-9246-4.ch011

Jain, S., Karampuri, R., & Chinthamalla, R. (2018). Solar PV Powered Standalone Water Pumping Systems. In P. Kumar, S. Singh, I. Ali, & T. Ustun (Eds.), *Handbook of Research on Power and Energy System Optimization* (pp. 139–190). IGI Global. doi:10.4018/978-1-5225-3935-3.ch005

Jana, S. K., Ghosh, M., & Karmakar, A. K. (2019). State-Level Status of Renewable Energy Development in India. In R. C. Das (Ed.), *Handbook of Research on Economic and Political Implications of Green Trading and Energy Use* (pp. 237–251). IGI Global. doi:10.4018/978-1-5225-8547-3.ch013

Jucha, S. O., & Moskal, G. J. (2018). Thermal Barrier Coatings on the Base of Samarium Zirconates. In A. H. Pakseresht (Ed.), *Production, Properties, and Applications of High Temperature Coatings* (pp. 79–106). IGI Global. doi:10.4018/978-1-5225-4194-3.ch004

Jumayev, A. Y., Jumayev, B. A., & Sariyev, K. A. (2018). Building Photoelectrical Power Station in the Region of Turkmenistan. In V. Kharchenko & P. Vasant (Eds.), *Handbook of Research on Renewable Energy and Electric Resources for Sustainable Rural Development* (pp. 61–85). IGI Global. doi:10.4018/978-1-5225-3867-7.ch003

Kalaiselvam, S., Rajan, D., & Ali, I. H. (2019). Thermal Technologies and Systems for Food Preservation. In P. D. Gaspar & P. D. da Silva (Eds.), *Novel Technologies and Systems for Food Preservation* (pp. 140–159). IGI Global. doi:10.4018/978-1-5225-7894-9.ch006

Kaliannan, J. B. A., Nhu, N. G., Dey, N., Ashour, A. S., & Balas, V. E. (2017). Automatic Generation Control of Hydro-Hydro Interconnected Power System Based on Ant Colony Optimization. In S. Patnaik (Ed.), *Recent Developments in Intelligent Nature-Inspired Computing* (pp. 153–170). IGI Global. doi:10.4018/978-1-5225-2322-2.ch007

Karkoodi, S., Aslani, A., Talebi, M., Roumi, S., & Abbassi, A. (2018). Transient 3D. *International Journal of Energy Optimization and Engineering*, *7*(3), 76–96. doi:10.4018/IJEOE.2018070105

Kaushik, S. C., Lamba, R., & Tyagi, S. K. (2018). Physics, Modelling, and Optimization Studies of Photon-Enhanced Thermionic Emission-Based Hybrid Energy Conversion System. In P. Kumar, S. Singh, I. Ali, & T. Ustun (Eds.), *Handbook of Research on Power and Energy System Optimization* (pp. 399–452). IGI Global. doi:10.4018/978-1-5225-3935-3.ch012

Keebler, D. W., Albertelli, P. D. Jr, & Mascarenhas, B. (2017). A Multi-Criteria Strategic Decision Making Model to Assess Renewable Energy Forms. *International Journal of Strategic Decision Sciences*, *8*(2), 1–10. doi:10.4018/IJSDS.2017040101

Keshavamurthy, R., Naveena, B. E., & Sekhar, N. (2018). Thermal Spray Coatings for Erosion-Corrosion Protection. In A. H. Pakseresht (Ed.), *Production, Properties, and Applications of High Temperature Coatings* (pp. 246–267). IGI Global. doi:10.4018/978-1-5225-4194-3.ch010

Khan, B., & Fanuel, M. (2020). Reliability Assessment of Microgrid-Integrated Electrical Distribution System. In B. Khan, H. H. Alhelou, & G. Hayek (Eds.), *Handbook of Research on New Solutions and Technologies in Electrical Distribution Networks* (pp. 136–156). IGI Global. doi:10.4018/978-1-7998-1230-2.ch008

Khanna, B. K. (2017). Vulnerability of the Lakshadweep Coral Islands in India and Strategies for Mitigating Climate Change Impacts. In Ganpat, W. & Isaac, W. (Eds.), Environmental Sustainability and Climate Change Adaptation Strategies (pp. 64-96). IGI Global. http://doi:10.4018/978-1-5225-1607-1.ch003

Kharchenko, V., Gusarov, V., & Bolshev, V. (2019). Reliable Electricity Generation in RES-Based Microgrids. In H. H. Alhelou & G. Hayek (Eds.), *Handbook of Research on Smart Power System Operation and Control* (pp. 162–187). IGI Global. doi:10.4018/978-1-5225-8030-0.ch006

Kharchenko, V., Panchenko, V., Tikhonov, P. V., & Vasant, P. (2018). Cogenerative PV Thermal Modules of Different Design for Autonomous Heat and Electricity Supply. In V. Kharchenko & P. Vasant (Eds.), *Handbook of Research on Renewable Energy and Electric Resources for Sustainable Rural Development* (pp. 86–119). IGI Global. doi:10.4018/978-1-5225-3867-7.ch004

Kirilov, L., Guliashki, V., & Staykov, B. (2019). Web-Based Decision Support System for Solving Multiple-Objective Decision-Making Problems. In N. Dey (Ed.), *Technological Innovations in Knowledge Management and Decision Support* (pp. 150–175). IGI Global. doi:10.4018/978-1-5225-6164-4.ch007

Koker, A. & Okur, Ä. Ozturkoglu-Budak, S. & Alpas, H. (2020). Non-Thermal Preservation of Dairy Products. In YÄ±kmÄ±ÅŸ, S. (Ed.), Technological Developments in Food Preservation, Processing, and Storage (pp. 1-25). IGI Global. http://doi:10.4018/978-1-7998-1924-0.ch001

Krim, S., Gdaim, S., Mtibaa, A., & Mimouni, M. F. (2019). Design of a Low-Cost Controller for Sun-Tracking to Maximize the Output of Photovoltaic Panels. *International Journal of Energy Optimization and Engineering*, *8*(1), 1–24. doi:10.4018/IJEOE.2019010101

Krishnamoorthy, S. & M. D. (2020). Photocatalytic Activity of ZnO Nanoparticles on the Degradation of Organic Dyes Under Solar Light. In *Handbook of Research on Emerging Developments and Environmental Impacts of Ecological Chemistry* (pp. 409-431). IGI Global. http://doi:10.4018/978-1-7998-1241-8.ch019

Kuada, J., Mensah, E., Bujac, A. I., & Bentzen, J. S. (2019). Marketing Solar Energy in Ghana. In A. Gbadamosi (Ed.), *Exploring the Dynamics of Consumerism in Developing Nations* (pp. 301–320). IGI Global. doi:10.4018/978-1-5225-7906-9.ch014

Kumar, P., & Thanki, D. V. (2018). Synchronized Operation of Grid Power, Solar Power, and Battery for Smart Energy Management. In P. Kumar, S. Singh, I. Ali, & T. Ustun (Eds.), *Handbook of Research on Power and Energy System Optimization* (pp. 255–291). IGI Global. doi:10.4018/978-1-5225-3935-3.ch008

Kumar, V. N., Panda, G., & Naidu, B. R. (2018). Seamless Control and Unified Dynamic Energy Management in a Renewable/Clean Energy Integrated Self-Reliant DC Microgrid. In P. Kumar, S. Singh, I. Ali, & T. Ustun (Eds.), *Handbook of Research on Power and Energy System Optimization* (pp. 511–551). IGI Global. doi:10.4018/978-1-5225-3935-3.ch015

Lacorde, M. (2019). Assessing the Environmental Characteristics of the Margaret River Wine Region, Australia. *International Journal of Applied Geospatial Research, 10*(3), 1–24. doi:10.4018/IJAGR.2019070101

Lemba, I. T., Ferreira Dias, M., & Robaina, M. (2020). Renewable Energy Policy in Africa and Policy Options for Sustainability. In R. Das & N. Mandal (Eds.), *Interdisciplinary Approaches to Public Policy and Sustainability* (pp. 153–176). IGI Global. doi:10.4018/978-1-7998-0315-7.ch008

Li, H., & Liu, Z. (2018). Performance Prediction and Optimization of Solar Water Heater via a Knowledge-Based Machine Learning Method. In P. Kumar, S. Singh, I. Ali, & T. Ustun (Eds.), *Handbook of Research on Power and Energy System Optimization* (pp. 55–74). IGI Global. doi:10.4018/978-1-5225-3935-3.ch002

Liu, M., Horton, L., Kang, J., Kimmons, R. M., & Lee, J. (2017). Making Learning Fun. In B. Dubbels (Ed.), *Transforming Gaming and Computer Simulation Technologies across Industries* (pp. 130–152). IGI Global. doi:10.4018/978-1-5225-1817-4.ch008

Liu, M., Liu, Y., Liu, C., Wang, J., & Wu, M. (2017). An Efficient Spatiotemporal Approach for Moving Object Detection in Dynamic Scenes. *International Journal of Information Technology and Web Engineering, 12*(3), 62–73. doi:10.4018/IJITWE.2017070106

Liu, W., Shi, H., Ma, H., & Tsai, S. (2018). Improving Renewable Energy Source in Automotive Applications. In S. Tsai, B. Liu, & Y. Li (Eds.), *Green Production Strategies for Sustainability* (pp. 210–237). IGI Global. doi:10.4018/978-1-5225-3537-9.ch012

Lopez, C. Y., Bueno, J. D., Torres, I. Z., Lopez, M. L., Macias, A. H., & Ãlvarez, J. E. (2019). TiO2 Nanotubes Transformation Into 4nm Anatase Nanoparticles. *International Journal of Applied Nanotechnology Research, 4*(2), 26–44. doi:10.4018/IJANR.2019070102

Lu, J., & Lang, Y. (2019). A Research of Ecological Farmhouse Design Based on the Sustainable Development Needs. In K. Siu & Y. Wong (Eds.), *Practice and Progress in Social Design and Sustainability* (pp. 48–68). IGI Global. doi:10.4018/978-1-5225-4183-7.ch003

Luo, D. Ma, W. Wu, L. & Nakata, K. (2018). CO2 Reduction by Photocatalysis on TiO2. In Advanced Solid Catalysts for Renewable Energy Production (pp. 268-299). IGI Global. http://doi:10.4018/978-1-5225-3903-2.ch010

Majumder, P., & Saha, A. K. (2019). Identification of Most Significant Parameter of Impact of Climate Change and Urbanization on Operational Efficiency of Hydropower Plant. *International Journal of Energy Optimization and Engineering*, *8*(3), 43–68. doi:10.4018/IJEOE.2019070103

Makdisie, C., Haidar, B., & Alhelou, H. H. (2018). An Optimal Photovoltaic Conversion System for Future Smart Grids. In P. Kumar, S. Singh, I. Ali, & T. Ustun (Eds.), *Handbook of Research on Power and Energy System Optimization* (pp. 601–657). IGI Global. doi:10.4018/978-1-5225-3935-3.ch018

Mallek, M., Euchi, J., & Jerbi, Y. (2020). A Review on Optimization Modeling of Hybrid Energy Systems. In J. Euchi (Ed.), *Transportation, Logistics, and Supply Chain Management in Home Healthcare* (pp. 29–62). IGI Global. doi:10.4018/978-1-7998-0268-6.ch003

Maphanyane, J. G., Mapeo, R. B., & Akinola, M. O. (2018). Applications of Geospatial Research in Earth Resources. In J. G. Maphanyane, R. B. Mapeo, & M. O. Akinola (Eds.), *Handbook of Research on Geospatial Science and Technologies* (pp. 15–31). IGI Global. doi:10.4018/978-1-5225-3440-2.ch002

Messina, J., Suepa, T., Snapp, S., Olson, J., Nejadhashemi, A. P., Murray, S., Moore, N., Frake, A., Fan, P., & Adhikari, U. (2017). Food System Resilience and Sustainability in Cambodia. *International Journal of Applied Geospatial Research*, *8*(3), 53–75. doi:10.4018/ijagr.2017070104

Min, T. Y., Tyagunov, M. G., & Deriugina, G. M. (2020). Evaluation of the Effectiveness of the Use of Programs in the Design of Power Complexes Based on Renewable Energy Resources. In V. Kharchenko & P. Vasant (Eds.), *Handbook of Research on Energy-Saving Technologies for Environmentally-Friendly Agricultural Development* (pp. 28–59). IGI Global. doi:10.4018/978-1-5225-9420-8.ch002

Mohamed, R. G., Ebrahim, M. A., Bendary, F. M., & Osman, S. A. (2017). Transient Stability Enhancement for 20 MW PV Power Plant via Incremental Conductance Controller. *International Journal of System Dynamics Applications*, *6*(4), 102–123. doi:10.4018/IJSDA.2017100106

Morganti, P., & Garofalo, G. (2020). Renewable Energy and Economic Growth. In C. Silvestri, M. Piccarozzi, & B. Aquilani (Eds.), *Customer Satisfaction and Sustainability Initiatives in the Fourth Industrial Revolution* (pp. 63–78). IGI Global. doi:10.4018/978-1-7998-1419-1.ch004

Mukhamediyev, R. I., Kiseleva, S., Yakunin, K., Mustakayev, R., & Muhamedijeva, J. (2019). Tools and Methods of Information Support for the Development of Renewable Energy. In V. Kharchenko & P. Vasant (Eds.), *Renewable Energy and Power Supply Challenges for Rural Regions* (pp. 39–65). IGI Global. doi:10.4018/978-1-5225-9179-5.ch002

Mukherjee, A. & Mitra, P. (2018). Physical Characteristics of CBD Synthesized CdZnS Thin Films. In Emergent Research on Polymeric and Composite Materials (pp. 186-208). IGI Global. http://doi:10.4018/978-1-5225-3023-7.ch008

Mukherjee, S. (2017). India's Looming Power Crisis and the Way Forward. *International Journal of Sustainable Economies Management*, *6*(1), 64–81. doi:10.4018/ijsem.2017010104

Musti, K. S., Iileka, H., & Shidhika, F. (2020). Industry 4.0-Based Enterprise Information System for Demand-Side Management and Energy Efficiency. In N. Prakash & D. Prakash (Eds.), *Novel Approaches to Information Systems Design* (pp. 137–163). IGI Global. doi:10.4018/978-1-7998-2975-1.ch007

N. K. (2020). Recent Trends in Cold Storage Systems. In *Technological Developments in Food Preservation, Processing, and Storage* (pp. 26-43). IGI Global. http://doi:10.4018/978-1-7998-1924-0.ch002

Naresh, B., Singh, V. K., & Sharma, V. K. (2020). Design of RF Rectenna on Thin Film to Power Wearable Electronics. In V. K. Singh, R. Tiwari, V. Dubey, Z. Ali, & A. K. Singh (Eds.), *Design and Optimization of Sensors and Antennas for Wearable Devices* (pp. 25–39). IGI Global. doi:10.4018/978-1-5225-9683-7.ch003

Nazari, M. A., Aslani, A., & Ghasempour, R. (2018). Analysis of Solar Farm Site Selection Based on TOPSIS Approach. *International Journal of Social Ecology and Sustainable Development*, *9*(1), 12–25. doi:10.4018/IJSESD.2018010102

Nebdi, H. (2019). Space Weather and Link to Climate Change. In K. Kahime, M. El Hidan, O. El Hiba, D. Sereno, & L. Bounoua (Eds.), *Handbook of Research on Global Environmental Changes and Human Health* (pp. 1–20). IGI Global. doi:10.4018/978-1-5225-7775-1.ch001

Nefedova, L. V., Solovyev, A. A., & Popova, O. (2019). Renewable Energy Sources Development in Rural Areas of African Countries. In V. Kharchenko & P. Vasant (Eds.), *Renewable Energy and Power Supply Challenges for Rural Regions* (pp. 154–167). IGI Global. doi:10.4018/978-1-5225-9179-5.ch007

Nersesian, R. L. (2017). Incorporating Simulation in Project Scoping. In Y. Raydugin (Ed.), *Handbook of Research on Leveraging Risk and Uncertainties for Effective Project Management* (pp. 406–434). IGI Global. doi:10.4018/978-1-5225-1790-0.ch019

Nersesian, R. L., & McManus, J. (2018). Risk Inherent in Matching Unreliability With Uncertainty. In K. D. Strang, M. E. Korstanje, & N. Vajjhala (Eds.), *Research, Practices, and Innovations in Global Risk and Contingency Management* (pp. 73–97). IGI Global. doi:10.4018/978-1-5225-4754-9.ch005

Nersesian, R. L., & Strang, K. D. (2017). Feasibility Approaches to Reduce the Unreliability of Gas, Nuclear, Coal, Solar and Wind Electricity Production. *International Journal of Risk and Contingency Management*, *6*(1), 54–69. doi:10.4018/IJRCM.2017010104

Nesterenkov, P. A., & Nesterenkov, A. G. (2020). Performance Improvement Study of Linear Photovoltaic Systems With Concentration of Solar Radiation. In V. Kharchenko & P. Vasant (Eds.), *Handbook of Research on Energy-Saving Technologies for Environmentally-Friendly Agricultural Development* (pp. 164–190). IGI Global. doi:10.4018/978-1-5225-9420-8.ch007

Nesterenkov, P. A., Nesterenkova, L. A., & Nesterenkov, A. G. (2018). Cogeneration Solar Systems With Concentrators of Solar Radiation. In V. Kharchenko & P. Vasant (Eds.), *Handbook of Research on Renewable Energy and Electric Resources for Sustainable Rural Development* (pp. 230–254). IGI Global. doi:10.4018/978-1-5225-3867-7.ch010

Nogin, S. Monteiro, J. Melgar, S. G. Peyroteo, J. Mortal, A. Santos, C. M. Livramento, J. Cardoso, P. J. & Semi, Ã. £o, J. (2018). A Platform for the Promotion of Energy Efficiency and Monitoring in Hotel Units. In Handbook of Research on Technological Developments for Cultural Heritage and eTourism Applications (pp. 420-448). IGI Global. http://doi:10.4018/978-1-5225-2927-9.ch020

P. S. & C. S. (2020). Existing Issues Associated With Electric Distribution System. In *Handbook of Research on New Solutions and Technologies in Electrical Distribution Networks* (pp. 32-47). IGI Global. http://doi:10.4018/978-1-7998-1230-2.ch002

Panchenko, V., Chirskiy, S., & Kharchenko, V. V. (2020). Application of the Software System of Finite Element Analysis for the Simulation and Design Optimization of Solar Photovoltaic Thermal Modules. In V. Kharchenko & P. Vasant (Eds.), *Handbook of Research on Smart Computing for Renewable Energy and Agro-Engineering* (pp. 106–131). IGI Global. doi:10.4018/978-1-7998-1216-6.ch005

Panchenko, V., & Kharchenko, V. (2020). Development and Research of PVT Modules in Computer-Aided Design and Finite Element Analysis Systems. In V. Kharchenko & P. Vasant (Eds.), *Handbook of Research on Energy-Saving Technologies for Environmentally-Friendly Agricultural Development* (pp. 314–342). IGI Global. doi:10.4018/978-1-5225-9420-8.ch013

Pandey, A., Tiwari, S. P., Dutta, V., & Kumar, V. (2018). Photon Upconversion in Lanthanide-Activated Inorganic Luminescent Materials. In R. Tiwari, V. Dubey, & S. J. Dhoble (Eds.), *Emerging Synthesis Techniques for Luminescent Materials* (pp. 86–116). IGI Global. doi:10.4018/978-1-5225-5170-6.ch004

Pathare, P. B., Roskilly, A. P., & Jagtap, S. (2019). Energy Efficiency in Meat Processing. In P. D. Gaspar & P. D. da Silva (Eds.), *Novel Technologies and Systems for Food Preservation* (pp. 78–107). IGI Global. doi:10.4018/978-1-5225-7894-9.ch004

Pelau, C., Pop, M., & Pop, A. (2019). Comparative Analysis Regarding the Online Interest of Consumers Towards Green Retrofitting. In A. Tantau (Ed.), *Retrofitting for Optimal Energy Performance* (pp. 73–95). IGI Global. doi:10.4018/978-1-5225-9104-7.ch004

Pramanik, P. K., Mukherjee, B., Pal, S., Pal, T., & Singh, S. P. (2020). Green Smart Building. In A. Solanki & A. Nayyar (Eds.), *Green Building Management and Smart Automation* (pp. 1–50). IGI Global. doi:10.4018/978-1-5225-9754-4.ch001

Price, J. Abel, M. O. Varney, A. & Wexler, D. (2019). Positive Energy. In Handbook of Research on Science Literacy Integration in Classroom Environments (pp. 203-219). IGI Global. http://doi:10.4018/978-1-5225-6364-8.ch013

Prusov, V., & Doroshenko, A. (2018). *Brief Description of the Main Characteristics of the Earth-Atmosphere System*. IGI Global. doi:10.4018/978-1-5225-2636-0.ch001

Radhakrishnan, S., & Vijayarajan, V. (2019). Farming 4.0. In K. Saravanan, G. Julie, & H. Robinson (Eds.), *Handbook of Research on Implementation and Deployment of IoT Projects in Smart Cities* (pp. 279–295). IGI Global. doi:10.4018/978-1-5225-9199-3.ch017

Rao, P. G., Jagadeesh, V., Santa Rao, K., & Chittibabu, V. (2017). Performance of Solar Power Refrigerator Using Different Materials. *International Journal of Manufacturing, Materials, and Mechanical Engineering*, *7*(1), 52–66. doi:10.4018/IJMMME.2017010104

Razi, M., & Ali, Y. (2019). Social, Environmental and Economic Impacts of Adopting Clean Energy Use. *International Journal of Decision Support System Technology, 11*(4), 29–49. doi:10.4018/IJDSST.2019100102

Rehman, A. U., Javid, T., Khan, I. U., & Murtaza, A. (2017). Thermal Imaging in Smart Applications. In V. Santhi (Ed.), *Recent Advances in Applied Thermal Imaging for Industrial Applications* (pp. 147–174). IGI Global. doi:10.4018/978-1-5225-2423-6.ch006

Rwegasira, D. S., Ben Dhaou, I. S., Kondoro, A., Anagnostou, A., Kelati, A., Naiman, S., Taylor, S. J., Mvungi, N. H., & Tenhunen, H. (2019). A Demand-Response Scheme Using Multi-Agent System for Smart DC Microgrid. *International Journal of Embedded and Real-Time Communication Systems*, *10*(1), 48–68. doi:10.4018/IJERTCS.2019010103

Sajeva, M., Mitchell, A., & Lemon, M. (2019). Reconfiguring Household Management in Times of Discontinuity as an Open System. *International Journal of Food and Beverage Manufacturing and Business Models*, *4*(1), 1–19. doi:10.4018/IJFBMBM.2019010101

Sakhlecha, M., Bajpai, S., & Singh, R. K. (2019). Evaluating the Environmental Impact Score of a Residential Building Using Life Cycle Assessment. *International Journal of Social Ecology and Sustainable Development, 10*(4), 1–16. doi:10.4018/IJSESD.2019100101

Salima, M., Loubna, A., & Riad, T. (2018). A Global Stability of Linearizing Control of Induction Motor for PV Water Pumping Application. *International Journal of System Dynamics Applications, 7*(3), 31–56. doi:10.4018/IJSDA.2018070102

Sambhanthan, A. (2019). A Green Energy Management Framework for Software Development Firms. In B. Christiansen, I. Sysoeva, A. Udovikina, & A. Ketova (Eds.), *Emerging Economic Models for Global Sustainability and Social Development* (pp. 312–328). IGI Global. doi:10.4018/978-1-5225-5787-6.ch017

Sen, V., & Kulkarni, A. (2020). Promotional Policies and Legislative Support for Grid-Connected Renewable Energy Projects. In P. Yang (Ed.), *Cases on Green Energy and Sustainable Development* (pp. 60–94). IGI Global. doi:10.4018/978-1-5225-8559-6.ch003

Sennouni, M. A., Abboud, B., Tribak, A., Bennis, H., & Latrach, M. (2017). Advance and Innovation in Wireless Power Transmission Technology for Autonomous Systems. In A. El Oualkadi & J. Zbitou (Eds.), *Handbook of Research on Advanced Trends in Microwave and Communication Engineering* (pp. 316–361). IGI Global. doi:10.4018/978-1-5225-0773-4.ch010

Senshaw, D. A., & Edwards, A. (2020). Countries Progress in Solar PV in Support of NDC Implementation and Contribution to Achieving SDG7. In P. Yang (Ed.), *Cases on Green Energy and Sustainable Development* (pp. 1–29). IGI Global. doi:10.4018/978-1-5225-8559-6.ch001

Shahien, M. (2018). Reactive Plasma Spray. In A. H. Pakseresht (Ed.), *Production, Properties, and Applications of High Temperature Coatings* (pp. 299–332). IGI Global. doi:10.4018/978-1-5225-4194-3.ch012

Shajkumar, A., & Ramadoss, A. (2020). Recent Advancements in Photocatalytic Nanocomposites. In G. Clarizia & P. Bernardo (Eds.), *Diverse Applications of Organic-Inorganic Nanocomposites* (pp. 136–161). IGI Global. doi:10.4018/978-1-7998-1530-3.ch006

Shendryk, V., Boiko, O., Parfenenko, Y., Shendryk, S., & Tymchuk, S. (2019). Decision Making for Energy Management in Smart Grid. In V. G. Diaz & J. Bonilla (Eds.), *Handbook of Research on Industrial Advancement in Scientific Knowledge* (pp. 264–297). IGI Global. doi:10.4018/978-1-5225-7152-0.ch014

Shokhzoda, B. T., & Tyagunov, M. G. (2019). Solar Concentrators. In V. Kharchenko & P. Vasant (Eds.), *Renewable Energy and Power Supply Challenges for Rural Regions* (pp. 1–38). IGI Global. doi:10.4018/978-1-5225-9179-5.ch001

Shrestha, S., & Harrison, T. (2019). Using Machinima as Teaching and Learning Materials. *International Journal of Computer-Assisted Language Learning and Teaching*, *9*(2), 37–52. doi:10.4018/IJCALLT.2019040103

Simachew, B. (2020). Loss Minimization in Active Distribution Network. In B. Khan, H. H. Alhelou, & G. Hayek (Eds.), *Handbook of Research on New Solutions and Technologies in Electrical Distribution Networks* (pp. 119–135). IGI Global. doi:10.4018/978-1-7998-1230-2.ch007

Singh, B. P. & Gore, M. M. (2019). Smart DC Microgrid. In Technological Developments in Industry 4.0 for Business Applications (pp. 100-128). IGI Global. http://doi:10.4018/978-1-5225-4936-9.ch005

Singh, R. & Dutta, S. (2018). Visible Light Active Nanocomposites for Photocatalytic Applications. In Composites and Advanced Materials for Industrial Applications (pp. 270-296). IGI Global. http://doi:10.4018/978-1-5225-5216-1.ch012

Singh, R. S., & Abbod, M. (2020). Self-Assistive Controller Using Voltage Droop Method for DC Distributed Generators and Storages. In P. Yang (Ed.), *Cases on Green Energy and Sustainable Development* (pp. 379–405). IGI Global. doi:10.4018/978-1-5225-8559-6.ch014

Sivaraman, P., & Sharmeela, C. (2020). Solar Micro-Inverter. In J. Zbitou, C. I. Pruncu, & A. Errkik (Eds.), *Handbook of Research on Recent Developments in Electrical and Mechanical Engineering* (pp. 283–303). IGI Global. doi:10.4018/978-1-7998-0117-7.ch010

Skouri, S., & Bouadila, S. (2017). Optical Analysis of Solar Concentrators Using Photogrammetry Process. In V. Santhi (Ed.), *Recent Advances in Applied Thermal Imaging for Industrial Applications* (pp. 175–201). IGI Global. doi:10.4018/978-1-5225-2423-6.ch007

Solovyev, D. A., Morgunova, M. O., & Solovyev, A. A. (2019). Creating a Synergy. In V. Kharchenko & P. Vasant (Eds.), *Renewable Energy and Power Supply Challenges for Rural Regions* (pp. 136–153). IGI Global. doi:10.4018/978-1-5225-9179-5.ch006

Soltani, A., & Sharifi, E. (2019). Understanding and Analysing the Urban Heat Island (UHI) Effect in Micro-Scale. *International Journal of Social Ecology and Sustainable Development, 10*(2), 14–28. doi:10.4018/IJSESD.2019040102

Sony, M. (2019). A Survey on Household Energy Saving Measures for Last Decade in Western India. *International Journal of Public and Private Perspectives on Healthcare, Culture, and the Environment, 3*(1), 14–27. doi:10.4018/IJPPPHCE.2019010102

Soulayman, S. (2018). *Economic Consideration*. IGI Global. doi:10.4018/978-1-5225-2950-7.ch008

Soulayman, S. (2018). *Extraterrestrial Solar Radiation*. IGI Global. doi:10.4018/978-1-5225-2950-7.ch003

Soulayman, S. (2018). *Geographic Orientations*. IGI Global. doi:10.4018/978-1-5225-2950-7.ch002

Soulayman, S. (2018). *Optimum Tilt Angle Determine*. IGI Global. doi:10.4018/978-1-5225-2950-7.ch005

Soulayman, S. (2018). *Solar and Collector Angles*. IGI Global. doi:10.4018/978-1-5225-2950-7.ch001

Soulayman, S. (2018). *Solar Tracking*. IGI Global. doi:10.4018/978-1-5225-2950-7.ch006

Soulayman, S. (2018). *Technical Consideration*. IGI Global. doi:10.4018/978-1-5225-2950-7.ch007

Soulayman, S. (2018). *Terrestrial Solar Radiation*. IGI Global. doi:10.4018/978-1-5225-2950-7.ch004

Spataru, C. (2019). *A Global Perspective on Experiences and Practices for Low Carbon Technologies and Renewable Energy in Islands*. IGI Global. doi:10.4018/978-1-5225-6002-9.ch003

Strebkov, D., Nekrasov, A., Trubnikov, V., & Nekrasov, A. (2018). Single-Wire Resonant Electric Power Systems for Renewable-Based Electric Grid. In V. Kharchenko & P. Vasant (Eds.), *Handbook of Research on Renewable Energy and Electric Resources for Sustainable Rural Development* (pp. 449–474). IGI Global. doi:10.4018/978-1-5225-3867-7.ch019

Tantau, A., & Staiger, R. (2018). *Business Development and Environmental Impact in the Solar (PV) Field*. IGI Global. doi:10.4018/978-1-5225-2688-9.ch002

Tantau, A., & Staiger, R. (2018). *Business Models With H2*. IGI Global. doi:10.4018/978-1-5225-2688-9.ch003

Teoh, H. C. (2020). The Radio Sun and Planets. In K. H. Yeap & K. Hirasawa (Eds.), *Analyzing the Physics of Radio Telescopes and Radio Astronomy* (pp. 230–249). IGI Global. doi:10.4018/978-1-7998-2381-0.ch013

Tikhomirov, D., Vasiliev, A., & Dudin, S. (2019). Energy-Saving Electrical Installations for Heat Supply of Agricultural Objects. In V. Kharchenko & P. Vasant (Eds.), *Advanced Agro-Engineering Technologies for Rural Business Development* (pp. 96–122). IGI Global. doi:10.4018/978-1-5225-7573-3.ch004

Tikhonov, P. V., Mayorov, V. A., & Morenko, K. S. (2020). Energy-Saving Systems Using Photovoltaic Modules. In V. Kharchenko & P. Vasant (Eds.), *Handbook of Research on Smart Computing for Renewable Energy and Agro-Engineering* (pp. 464–485). IGI Global. doi:10.4018/978-1-7998-1216-6.ch018

Tomar, P. (2020). Role of Renewable Energy Techniques to Design and Develop Sustainable Green Building. In A. Solanki & A. Nayyar (Eds.), *Green Building Management and Smart Automation* (pp. 51–63). IGI Global. doi:10.4018/978-1-5225-9754-4.ch002

Tomor, Z. (2020). Citizens in the Smart City. *International Journal of Public Administration in the Digital Age*, *7*(1), 1–16. doi:10.4018/IJPADA.2020010101

Tsikaloudaki, K. Tsirigoti, D. Tsoka, S. & Theodosiou, T. (2020). Upgrading the Building Facades in Low-Density Residential Areas. In Handbook of Research on Urban-Rural Synergy Development Through Housing, Landscape, and Tourism (pp. 216-243). IGI Global. http://doi:10.4018/978-1-5225-9932-6.ch011

Udayakumar, E., Balamurugan, S., & Vetrivelan, P. (2019). A Neural Network-Based Automatic Crop Monitoring Robot for Agriculture. In D. Goyal, S. Balamurugan, S. Peng, & D. S. Jat (Eds.), *The IoT and the Next Revolutions Automating the World* (pp. 203–212). IGI Global. doi:10.4018/978-1-5225-9246-4.ch013

Vallejo-Fiallos, M. D. (2019). Corporate Social Entrepreneurship. In Handbook of Research on Ethics, Entrepreneurship, and Governance in Higher Education (pp. 305-331). IGI Global. http://doi:10.4018/978-1-5225-5837-8.ch014

Vaskov, A. G., Lin, Z. Y., Tyagunov, M. G., Shestopalova, T. A., & Deryugina, G. V. (2018). Design of Renewable Sources GIS for ASEAN Countries. In V. Kharchenko & P. Vasant (Eds.), *Handbook of Research on Renewable Energy and Electric Resources for Sustainable Rural Development* (pp. 1–25). IGI Global. doi:10.4018/978-1-5225-3867-7.ch001

Vaskov, A. G., Tyagunov, M. G., Shestopalova, T. A., Deryugina, G. V., & Ishchenko, I. (2018). Structure and Parameter Optimization of Renewable-Based Hybrid Power Complexes. In V. Kharchenko & P. Vasant (Eds.), *Handbook of Research on Renewable Energy and Electric Resources for Sustainable Rural Development* (pp. 352–382). IGI Global. doi:10.4018/978-1-5225-3867-7.ch015

Webb, H., & Liu, S. (2020). The Sustainable Falcon. In H. C. Webb & H. A. Al Numairy (Eds.), *Entrepreneurial Innovation and Economic Development in Dubai and Comparisons to Its Sister Cities* (pp. 1–16). IGI Global. doi:10.4018/978-1-5225-9377-5.ch001

Xiao, J., Cao, M., Li, X., & Hansen, P. (2020). Assessing the Effectiveness of the Augmented Reality Courseware for Starry Sky Exploration. *International Journal of Distance Education Technologies*, *18*(1), 19–35. doi:10.4018/IJDET.2020010102

Xie, W. (2018). Thermal Spray Methods and Splat Formation. In A. H. Pakseresht (Ed.), *Production, Properties, and Applications of High Temperature Coatings* (pp. 1–24). IGI Global. doi:10.4018/978-1-5225-4194-3.ch001

Yahyaoui, I., Tadeo, F., & Vieira Segatto, M. E. (2017). Smart Control Strategy for Small-Scale Photovoltaic Systems Connected to Single-Phase Grids. In M. A. Ferrag & A. Ahmim (Eds.), *Security Solutions and Applied Cryptography in Smart Grid Communications* (pp. 380–404). IGI Global. doi:10.4018/978-1-5225-1829-7.ch018

Yamaguchi, H. (2017). Supercritical Carbon Dioxide and Its Application to Rankine Cycle. In L. Chen & Y. Iwamoto (Eds.), *Advanced Applications of Supercritical Fluids in Energy Systems* (pp. 335–368). IGI Global. doi:10.4018/978-1-5225-2047-4.ch011

Yatimi, H., & Aroudam, E. (2018). Standalone Photovoltaic System with Maximum Power Point Tracking. *International Journal of System Dynamics Applications*, *7*(3), 94–111. doi:10.4018/IJSDA.2018070105

Yuferev, L., & Kuznetsov, P. N. (2019). Increasing the Energy Efficiency of Photovoltaic Systems Operating Under Conditions of Uneven Illumination. In V. Kharchenko & P. Vasant (Eds.), *Renewable Energy and Power Supply Challenges for Rural Regions* (pp. 88–114). IGI Global. doi:10.4018/978-1-5225-9179-5.ch004

Yülek, M. A., Bostanci, I., & Goksal, K. (2017). Data Center Regions and Economic Development. *International Journal of Social Ecology and Sustainable Development*, *8*(4), 32–47. doi:10.4018/IJSESD.2017100103

Zhu, L., Wu, M., Wan, X., Zhao, N., & Xiong, W. (2017). Image Recognition of Rapeseed Pests Based on Random Forest Classifier. *International Journal of Information Technology and Web Engineering*, *12*(3), 1–10. doi:10.4018/IJITWE.2017070101

Zouggar, S. T., & Adla, A. (2017). Proposal for Measuring Quality of Decision Trees Partition. *International Journal of Decision Support System Technology*, *9*(4), 16–36. doi:10.4018/IJDSST.2017100102

Zubov, D. (2017). Building IoT With Arduino. In P. Kocovic, R. Behringer, M. Ramachandran, & R. Mihajlovic (Eds.), *Emerging Trends and Applications of the Internet of Things* (pp. 117–166). IGI Global. doi:10.4018/978-1-5225-2437-3.ch005

About the Authors

Jagadish is an Assistant Professor at the Department of Mechanical Engineering, National Institute of Technology Raipur, Chhattisgarh India. He obtained his Ph.D. in Mechanical Engineering with specialization in Manufacturing Engineering from National Institute of Technology Silchar and Post graduation in Product Design and Development from National Institute of Technology Warangal, India. He has over 3 years of industrial experience in the field of design and analysis and 8+ years of teaching and research experience. He received an "Institutional Award (Gold Medal)" by Institution of Engineers India and "Best Innovative Award" by Springer for his outstanding research contribution and published more than 35 + research papers that includes SCI journals/Conference papers/Book & Book Chapters. His areas of interests are Green Manufacturing; Renewable Energy, Thermal Systems & its Optimization, MCDM methods, Advanced Manufacturing Process; Additive Manufacturing; Product Design & Development; Surface Texturing; CAD/CAM; Composite Materials; Decision Making Tools, etc.

Agnimitra Biswas is an Assistant Professor at the Department of Mechanical Engineering, National Institute of Technology Silchar, Assam, India. He obtained Ph.D. in Mechanical Engineering with specialization in Thermal Engineering from National Institute of Technology Silchar and Post graduation in Thermal Engineering from National Institute of Technology Silchar, India. His areas of interests are Renewable Energy, Thermal Engineering, Solar Energy, and Vertical Axis Wind Turbine. He has over 9 years of teaching and research experience. He has published published more than 45 research papers and 3 book chapters. Currently, he is doing research in solar energy collectors, water turbine, and Vertical axis wind turbine.

Index

www.ingramcontent.com/pod-product-compliance
Lightning Source LLC
LaVergne TN
LVHW081321300926
849161LV00008B/1845